ADVANCE PRAISE

for

An Inconveni-

"The problematic human/earth relatior ... and Jackson and Jensen's book makes a ... to assessing our situation and envisioning a way forw ... who has a nagging feeling that something is wrong and doesn't understand the breadth and depth of the problem or how to grapple with it should read this book." —Lisi Krall, author of *Proving Up*

"*An Inconvenient Apocalypse* pulls no punches. Wes Jackson and Robert Jensen, in this work of Anthropocenic soul-searching, offer an honest, accessible, and ruefully playful look at their own lives and at the predicament of human civilization during this century of upheaval and denial." —Scott Slovic, co-editor of *Ecoambiguity, Community, and Development*

"Wes Jackson and Bob Jensen have written *Common Sense* for our time. This book might be the spark that catalyzes the American Evolution." —Peter Buffett, co-president of the NoVo Foundation

"This is one of the most important books of our lifetime. *An Inconvenient Apocalypse* can help us face the difficult choices that confront us all and enable us to acknowledge the urgency of our current circumstance." —Frederick L. Kirschenmann, author of *Cultivating an Ecological Conscience*

"If you're already concerned about our species' survival prospects, this book will take you to the next level of understanding. Jackson and Jensen are clear and deeply moral thinkers, and their assessment of humanity's precarious status deserves to be widely read." —Richard Heinberg, author of *Power*

"In this essential contribution to the public debate, Wes Jackson and Robert Jensen critique the capitalist forces accelerating the climate crisis and the intellectual-activists who have balked at calling for the radical changes in human behavior that could mitigate, if not prevent, environmental and societal collapse. Their contribution will prove as enduring as it is timely." —Jason Brownlee, author of *Authoritarianism in an Age of Democratization*

"With intrepid honesty, tenderness, and grace, Jackson and Jensen lay out a clear framework for making sense of the most elusive complexities of climate crisis. Through kindred reflections and incisive analysis, they boldly enlighten readers of the probable and the possible in the decades to come. An affirmation and solace for the weary. A beacon for those seeking courage and understanding in unsettling times." —Selina Gallo-Cruz, author of *Political Invisibility and Mobilization*

"While making no religious claims, Jackson and Jensen engage the core questions that religious people must ask, if their own witness is to be credible: Who are we, and where are we in history? Do we have the capacity to make drastic change for the sake of a decent human future? Can we live with humility and grace instead of arrogance and an infatuation with knowledge devoid of wisdom? Read and consider." —Ellen F. Davis, author of *Scripture, Culture, and Agriculture*

"The nature of all living organisms, so this book argues, is to go after 'dense energy,' resulting eventually in crisis. If that is so, then the human organism is facing a tough question: Can we overcome our own nature? Courageous and humble, bold and provocative, the authors of *An Inconvenient Apocalypse* do not settle for superficial answers." —Donald Worster, author of *Shrinking the Earth*

AN INCONVENIENT APOCALYPSE

This book was selected as the 2022 Giles Family Fund Recipient. The University of Notre Dame Press and the author thank the Giles family for their generous support.

Giles Family Fund Recipients

2019 *The Glory and the Burden*, Robert Schmuhl (expanded edition, 2022)

2020 *Ars Vitae: The Fate of Inwardness and the Return of the Ancient Arts of Living*, Elisabeth Lasch-Quinn

2021 *William Still: The Underground Railroad and the Angel at Philadelphia*, William C. Kashatus

2022 *An Inconvenient Apocalypse: Environmental Collapse, Climate Crisis, and the Fate of Humanity*, Wes Jackson and Robert Jensen

The Giles Family Fund supports the work and mission of the University of Notre Dame Press to publish books that engage the most enduring questions of our time. Each year the endowment helps underwrite the publication and promotion of a book that sparks intellectual exploration and expands the reach and impact of the university.

AN INCONVENIENT APOCALYPSE

Environmental Collapse, Climate Crisis, and the Fate of Humanity

WES JACKSON AND **ROBERT JENSEN**

University of Notre Dame Press
Notre Dame, Indiana

Notre Dame, Indiana 46556
undpress.nd.edu

Published in the United States of America

Library of Congress Control Number: 2022935758

ISBN: 978-0-268-20365-8 (Hardback)
ISBN: 978-0-268-20366-5 (Paperback)
ISBN: 978-0-268-20367-2 (WebPDF)
ISBN: 978-0-268-20364-1 (Epub)

To Jack Ewel, a first-rate ecologist whose rigorous research
and kindly collegiality have always been a standard

CONTENTS

ACKNOWLEDGMENTS

Thanks to colleagues and comrades from The Land Institute, Ecosphere Studies network, and New Perennials Project for spirited intellectual engagement.

readers to bear with us while we explain our sense of urgency, our approach to analyzing the crises, and what we believe are our best options for the future. We offer this advice on intellectual engagement from John Steinbeck:

> We had had many discussions at the galley table and there had been many honest attempts to understand each other's thinking. There are several kinds of reception possible. There is the mind which lies in wait with traps for flaws, so set that it may miss, though not grasping it, a soundness. There is a second which is not reception at all, but blind flight because of laziness, or because some pattern is disturbed by the processes of the discussion. The best reception of all is that which is easy and relaxed, which says in effect, "Let me absorb this thing. Let me try to understand it without private barriers. When I have understood what you are saying, only then will I subject it to my own scrutiny and my own criticism." This is the finest of all critical approaches and the rarest.[1]

We do not claim our arguments are flawless, and we hope readers who identify a flaw will not dismiss immediately the soundness of the larger analysis. We will suggest that there are patterns at work in the world quite different from what many people believe. We encourage readers to scrutinize and criticize after engaging with our arguments in the easy and relaxed fashion advised by Steinbeck. We agree that such a critical approach is rare, because we can reflect on how often we have taken the other two routes in our lives. As we get older, we hope we have gotten better at being relaxed.

Contemporary Crises

There are many different ways to categorize and analyze the threats that humans have created and cannot evade much longer. Reasonable people can disagree about details, but for the moment let's focus on the big picture. The following summary would get wide agreement not only in progressive political circles but also in much of mainstream society and even in parts of more conservative political communities.

First, within the human family, we face a struggle for social justice in societies that currently do not operate in a manner consistent with widely held values concerning dignity, solidarity, and equality. Many people, whatever their political affiliation, express a commitment to (1) the inherent dignity of all people, (2) the importance of solidarity for healthy community life, and (3) the need for a level of equality that makes dignity and solidarity possible. But living according to those moral principles is hard, and the impediments are well known: the sexual and social subordination of women and girls under patriarchy; the brutal history and corrosive contemporary practices of white supremacy; wealth inequality and deprivation in unjust economic systems, most recently capitalism; and global inequality rooted in historical colonialism and today's economic imperialism. If we can't align our dysfunctional politics and destructive economics with those widely held values, we are in trouble.

Second, we face a struggle for an ecologically sustainable relationship between humans and the larger living world, the ecosphere.[2] That means dramatic changes—in both the way we think and the way we live—are necessary in societies that draw down the ecological capital of the ecosphere beyond replacement levels. This will require a shift away from the widely accepted idea that Earth exists for humans to exploit without much regard either for other organisms or for the long-term health of the ecosystems that sustain us. The threats here also are well known: human-centric economic systems and cultural norms that have for millennia led to soil erosion and degradation, and in recent centuries have produced chemical contamination of land and water, a steep decline in biodiversity, and climate destabilization. If we can't align our living arrangements with the laws of physics and chemistry, we are in trouble.

From those basics on which many agree, disagreements emerge pretty quickly about details, even among like-minded people. These conflicts often are the result of different core ideologies or different locations in the social structure. But we think this is a reasonable general summary of the challenges we face in achieving social justice and ecological sustainability.

How are we to proceed? We should consider possible actions in the context of *multiple cascading crises*[3]—three words we would like to etch into everyone's consciousness. The concurrent and unpredictable failures of social systems produce short-term threats that demand a response immediately, as well as long-term threats that will not be resolved anytime soon and perhaps cannot be resolved. We do not face discrete problems that can be solved in isolation. We have to struggle to understand all of these destructive forces and how they interact so that our actions will be as effective as possible. All of these threats demand everyone's immediate attention, yet no one person can act on all fronts in any given moment or even in a lifetime. In view of all that, it is not surprising that many people find it easy to lose hope that the necessary change will happen in time.

Abandoning such hope is reasonable, an assertion to which we return at the end of the book. But for now, we want to point out that there's no sense pretending these threats are not overwhelming simply because we wish they were more manageable. It's fine in a love song to suggest that "the impossible will take a little while,"[4] but in real life the impossible is just that, impossible. Where does that leave us? The task is to recognize what is impossible and what is—or at least might be—possible, not in the abstraction of theory, but in the concreteness of the world.

The coming decades are likely to be marked by dramatic dislocations as a result of our social and ecological crises. We say "likely," because no one can predict with precision next week's weather, let alone the exact trajectory of human societies in the coming decades. Still, we believe that feeling some despair in the face of these threats is a rational, reasonable, and responsible reaction. Such despair—not over our personal fates but our species' collective inability to value the larger living world—should be pondered, not waved away with platitudes. We do not advocate nihilism, but we take seriously the biophysical limits of the ecosphere and human limits. It's cowardly to say that nothing can be done. It's silly to say that we can do whatever we set our minds to. If we stop fantasizing about doing the impossible, we can focus on doing the best job we can to achieve what is possible. The two of us continue to spend a considerable amount of our own time focused on analyzing and responding to these threats, well aware that positive outcomes are not guaranteed. Analysis (no matter how grim) and action (no matter how slim the chances of success) are antidotes to despair, though we are not pretending it is simple to determine what is possible and what is impossible, or what the most effective personal and policy choices might be.

So here's another thing that needs to be right up front: it is highly unlikely that the destructive forces unleashed by humans over the past ten thousand years since the invention of agriculture will be stopped in time to avoid what might be called apocalyptic consequences. Don't worry, that's not a lead-in to rapture talk. We use the term "apocalyptic" in a secular sense, as we explain in chapter 3. Because we remain engaged in the struggle and hope others will too, we want to suggest another way to approach crises, a way to organize our thinking that sharpens rather than obscures our understanding of the impediments to the deep changes necessary.

We come back to those three crucial words—multiple cascading crises—that remind us that the task is not to solve separate problems but to see the failure of systems and recognize the limits of our ability to predict how those failures will play out over time. Given the scope of that task, we are going to need what the geologist T. C. Chamberlin called multiple working hypotheses. In an essay first published in 1897, Chamberlin warned that a single "ruling theory" can too easily lead us to ignore evidence that disproves the conventional wisdom or supports an alternate explanation.[5] In our context, that means we should beware anyone selling an easy way out, a way for us to have it all without dramatic changes.

If we are to respond creatively and effectively, we also need to think at both the material and ideological levels—the tangible and the intangible, the way we live day to day, and the way we think about the meaning of our lives. At the material level, we face a *crisis of consumption*. In aggregate terms, the human population has too much stuff. That stuff is not equally or equitably distributed among the population, of course. But no matter the level of fairness and justice in societies, the ecological costs of the extraction, processing, and waste disposal required

all, it's mostly white guys, young and old, who got us into the various messes we're in.

On confidence: In our combined 150 years (when this book is published, Jackson will be 86 and Jensen, 64), we have studied in a variety of disciplines and intellectual traditions, and we have been active in a variety of political and social movements. We have spent considerable time and intellectual energy pondering pressing questions.

We each have a contrarian streak, but we aren't writing in search of controversy for its own sake. We don't critique conservatives to appeal to liberals, nor do we challenge progressive ideology just to shore up our credentials with the mainstream. We fit somewhere on the left side of the political fence, so we have learned to expect the conservative resistance that comes our way. But we also challenge progressive assumptions and know from experience that some of these conclusions will be rejected on the Left. We also are supporters of modern science, so we expect pushback from the critics of that tradition. But we also make claims that challenge Enlightenment dogma.

We perhaps have gone on too long here, but we want to make it clear that this book is born out of a sense of responsibility, not entitlement, out of a sense of urgency to face the multiple cascading crises. For a good chunk of our adult lives, both of us have been subsidized to do intellectual work—defined here simply as collecting and assessing evidence to try to identify patterns in the world, the attempt to understand how both physical and social systems work. We've had access to good schools, first-class libraries, and conversation with lots of smart people all over the world, which have informed our teaching, writing, and activism. We're grateful for everything those things have brought to our lives and believe we should make responsible use of those advantages.

And here's the irony. We have a lot of running room to be as blunt as necessary because we are financially stable and don't have to worry about losing our jobs, and we are old enough that we don't much care about risking whatever status we've acquired over the years. As T. S. Eliot put it in "East Coker," "Old men ought to be explorers." We take seriously Eliot's charge that we "must be still and still moving / Into another intensity."[6] We can say what we think needs to be said, especially when it challenges the positions taken by friends and colleagues who may bristle at our intensity. If we thought the analyses of the political and intellectual movements of which we are a part were 100 percent correct, we would have no reason to write. We wish that were the case, but it's not.

What We Believe

We aren't going to march readers through the volumes of research demonstrating the depth of the social disparities and the severity of the ecological dangers. A parade of statistics and studies rarely persuades those who have decided to ignore the threats to human communities and ecosystems. This book takes those disparities and dangers as a starting point.

To anchor us in the depth of those threats, consider this warning from 1,700 of the world's leading scientists:

> Human beings and the natural world are on a collision course. Human activities inflict harsh and often irreversible damage on the environment and on critical resources. If not checked, many of our current practices put at serious risk the future that we wish for human society and the plant and animal kingdoms, and may so alter the living world that it will be unable to sustain life in the manner that we know. Fundamental changes are urgent if we are to avoid the collision our present course will bring about.[7]

That statement was issued in 1992. Because such concerns were ignored by the people in power, in 2017 a follow-up report, "World Scientists' Warning to Humanity: A Second Notice," was issued. This time, more than 15,000 scientists from 184 countries made the case: "We are jeopardizing our future by not reining in our intense but geographically and demographically uneven material consumption and by not perceiving continued rapid population growth as a primary driver behind many ecological and even societal threats."[8]

For a quick summary of the threats, a 2020 report from the Commission for the Human Future, a nongovernmental and nonpartisan group of researchers and citizens, identified ten catastrophic risks that require immediate attention.

- The decline of key natural resources and an emerging global resource crisis, especially in water;
- The collapse of ecosystems that support life, and the mass extinction of species;
- Human population growth and demand, beyond the Earth's carrying capacity;
- Global warming, sea-level rise, and changes in the Earth's climate affecting all human activity;
- Universal pollution of the Earth system and all life by chemicals;
- Rising food insecurity and failing nutritional quality;
- Nuclear arms and other weapons of mass destruction;
- Pandemics of new and untreatable disease;
- The advent of powerful, uncontrolled new technologies;
- National and global failure to understand and act preventively on these risks.[9]

The implications of these warnings, which so far have gone largely unheeded not only by those in power but also by the dominant culture more generally, are clear. If we don't transcend a growth economy, there are hard times ahead. And if we do manage to construct a new economic order, there are hard times ahead. Hard times are coming for everyone, even though some people are more responsible for social and ecological problems than others and some of those people will be able to evade the consequences of those problems, at least in the short term.

The task of creating new systems is daunting, in large part because of challenges posed by the nature of the human animal, combined with most people's denial of what it means to be an animal. We have faith in the better angels of our nature but realize that those better angels alone won't save us from what we call "the temptations of dense energy," which have come most recently in the form of fossil fuels. We conclude that there are no workable solutions to the most pressing problems of our historical moment. The best we can do is minimize the suffering and destruction. We conclude that the human future—even if today's progressive social movements were to be as successful as possible—will be gritty and grim. And yet we continue to search for ways to make whatever changes are possible, not just out of moral obligation, but in the pursuit of a more joyful participation in the Creation.

We like the term "Creation" for the sense of reverence that it brings, even though we don't believe in a creator god. But we see creativity and ongoing creation all around us, every day, in the ecosphere. We have affection and respect for religious traditions but are critical of the dogma that those traditions so often produce. And we get just as nervous when secular traditions drift into dogma, as they regularly do. We work hard to monitor our own thinking to avoid overdriving our headlights, well aware that the human weakness of claiming certainty in the face of complexity lies in wait for us just as much as for others.

No matter what their identity, all authors should ask before writing, "Is another book really necessary?" We ask ourselves, "Is another book *by us* necessary?" Our answer for this project is obviously "Yes," because many people are looking for an honest account of the human condition at this tenuous moment in history, one that does not give up on the obligation to act but also does not turn away from the inevitable grief when we confront our limitations. We believe that the claims we make in this book—especially the ones that might make some people bristle—are not only defensible, but crucial to consider if there is to be a decent human future, perhaps if there is to be a human future at all.

Same and Different

Right up front, we made it clear that self-reflection about identity is important. Most people today accept that things like sex, race, class, and national status are relevant in shaping people's ideas. But within those categories, there is considerable individual variation. In the interest of full disclosure, a bit more about Jackson and Jensen, not simply in terms of identity categories, but individual biographies. While we share a similar social position, like any two people we are not the same. We

ONE

WHO IS "WE"?

The bigger "we"—the focus of our analysis in this book—is all of us, the entire human population, every human who has ever lived. In our introductions, we emphasized that humans today live—and throughout the past ten thousand years since the advent of agriculture have lived—with different levels of access to wealth and material comfort. That inequality is the product of differences in status and power of individuals within a society, as well as differences between societies. In this chapter we want to talk about humans as a species, in a universal sense. This kind of analysis is too often ignored out of a legitimate concern about overgeneralizing and obscuring important differences among people and societies, both historically and today. But understanding the universal "we" is a crucial part of the story of how the species got to a place where there could be such dramatic differences in life outcomes among individuals and between societies.

We start with a set of simple observations about sameness and difference. From one perspective, Jackson and Jensen are pretty much the same kind of person. In the big scheme of things, the differences between us are minor, even trivial. From another perspective, Jackson and Jensen are distinctly different in ways that really matter in determining how we show up in the world. Both those statements are true, not only of Jackson and Jensen, but of any two people on Earth. We all belong to the same species, with only minor variations. And each one of us is a truly distinctive individual. Even identical twins can be dramatically different from one another.

That may seem obvious, but both things are true and both things are relevant to understanding the challenges of being human at this historical moment. This isn't about whether particular societies are more collectivist or more individualistic in philosophy and policy. These observations about sameness and difference aren't a product of cultural differences. In every society that ever existed—no matter how a given society organizes and justifies its political and economic systems—both of these things are true: we are basically all the same animal, and we are all distinctly different individuals. It's tempting to offer the often-quoted F. Scott Fitzgerald observation that "the test of a first-rate intelligence is the ability to hold two opposed ideas in the mind at the same time, and still retain the ability to function."[1] But these aren't opposed ideas, just two aspects of reality that, at first glance, might seem to be at odds.

We take seriously this insight from Richard Levins and Richard Lewontin: "Things are similar: this makes science possible. Things are different: this makes science necessary."[2] Our interest in grappling with similarities and differences, in pondering the particular and the universal, is not to promote platitudes about world harmony and loving one another, though we believe harmony and love are good things. Instead, we want to (1) understand the forces that brought the species to this moment in history, in order to (2) better guide the choices we make today and in the future.

Understanding that there is a universal "we"—that we are all the same kind of animal—matters to our chances for maintaining a large-scale human presence on Earth as we figure out how to deal with increasingly limited options for future generations. Thinking through "we" is essential to human survival.[3]

Assigning Responsibility

The folksinger-storyteller Utah Phillips is often quoted by people in the environmental movement: "The Earth is not dying, it is being killed, and those who are killing it have names and addresses." Earth isn't literally being killed, of course. As many have pointed out, whatever damage humans do—and that damage can be considerable—we won't kill off all life or render the planet incapable of sustaining life. But Phillips's quip is a reminder of the point we will continue to emphasize: wealth and power, along with the responsibility for ecosystem degradation, are not distributed uniformly in the world. Some people take more and therefore should be more accountable for the effects of their taking.

But uncomfortable questions nag at us. First, how long is the list of people who are if not literally killing the Earth, significantly reducing the options of future generations? Are the addresses of Jackson and Jensen on that list? As much as we hate to acknowledge it, they are. Yours are likely there as well. Those of us in the more affluent sectors of the world—not just the 1 percent, but many, maybe most, of the remaining 99 percent of the developed world—are living at unsustainable levels for the current world population. Why would we exempt ourselves from accountability?

A list of examples of our complicity in ecological destruction would require a separate chapter. Just by living "on the grid" in the affluent industrial developed first world, we are in some sense contributing to the destruction of ecosystems, as is likely everyone reading this book. Cars are a part of the ecosphere-degrading project, for example. Depending on where one lives and works, it can be difficult, if not impossible, to function "normally" without a car. Still, is not every driver contributing to reducing the options of future generations?

This complicity does not, however, lead us to focus primarily on individuals' personal choices to change lifestyles, though we believe that a more frugal lifestyle is a good thing, both for people and for the ecosphere. Like most in the progressive environmental movement, we believe that the economic system and social norms have to change if we are to have a chance to create sustainable societies, and a focus only on individual choices within an unsustainable system ends up being counterproductive. Driving less—walking, biking, taking the bus—is a good thing, but it won't change the world if the larger car culture endures and cities are designed in such a way that living and working without a car is difficult. If you think trading in your gas-burning car for an electric vehicle is the solution, think again. The energy consumption and resource extraction required to manufacture EVs make them an ecologically unsustainable choice as well.[4]

We agree that the slogan of one network of ecosocialists, "System change not climate change,"[5] is a good place to start (more on this below). But once we get on board with system change, nagging questions remain. If system change should come tomorrow—if capitalism were replaced by an egalitarian economic system focused not on endless growth and profit but on people's needs—how easy would it be for everyone to give up most of the comforts to which we have grown accustomed, comforts that are directly implicated in ecosphere degradation? Who among us has never felt the temptations of dense energy—the ways that such energy can reduce the need for human labor and increase the pleasure that comes from more manufactured goods? Does holding the correct environmental position on system change and contributing to environmental organizing mean that we need not worry about those temptations? We believe that acknowledging the power of those temptations increases the potential effectiveness of that organizing.

The two of us have spent decades writing, teaching, and agitating in movements for a more just and sustainable system. We believe that work has been important. But that does not mean we find it easy to reject the benefits of dense energy, nor does it get us off the hook for our own unsustainable consumption.

These considerations lead us to a simple point: moral judgments of human failures are inevitable and necessary as we all work to hold ourselves and each other accountable. But in the work of creating a just and sustainable society, *the moral high ground is a dangerous place to stand*, even when standing there might be warranted, even when we have good reason to believe we have adopted the morally compelling political position for justifiable reasons, even when we are active in movements for change, even when we need to make necessary judgments about the failures of others. We'll say more about the dangers of the moral high ground later. For now, we want to make sure no one—including the authors—is feeling smug.

"We" Is All of Us

Let's sum up the state of affairs at his moment in history. *We* humans have made a mess of things, which is readily evident if *we* face the avalanche of studies and statistics describing the contemporary ecological crises *we* face. But even with the mounting evidence of the consequences for people and the ecosphere, *we* have not committed to a serious project to slow the damage that *we* do. Those who have little or no access to wealth and power would be within their rights to object, on the grounds that the "we" diffuses responsibility. Who has made a mess of things and who has failed to act? Who is to blame for the problems, and who is responsible for the costs? Put more bluntly, borrowing from the imagined exchange between the Lone Ranger and Tonto when they were in a tough fight with Indians, "What do you mean, *we*, white man?"

Our thesis is that while not every individual or culture is equally culpable, the human failure over the past ten thousand years is the result of the imperative of all life to seek out energy-rich carbon. Humans play that energy-seeking game armed with an expansive cognitive capacity and a species propensity to cooperate and develop a complex division of labor.[6] That's a way of saying that humans are smart and know how to coordinate our activities to leverage that advantage. Specific individuals and societies are morally accountable for their failures, and certain political and economic systems are central to those failures. But the failures are also the result of the kind of organisms we are. Both things are true, and both things are relevant.

The global North—which is to say, fossil fuel–powered capitalism as it developed in Europe—bears primary responsibility for the shape of the contemporary crises, and those societies have failed to meet their obligation or, in some cases, even acknowledge an obligation to change course. In our lifetimes, the primary force behind that failure has been the United States. Within affluent societies, the wealthy and powerful bear the greatest responsibility for destructive policies. But if there is to be a decent human future, we have to realize that human-carbon nature is at the core of the problem, a reality that exempts no one. We cannot ignore the relevance of "we."

This may sound harsh in a world with so much human suffering, so unequally distributed. So let us be clear. This analysis does not minimize or trivialize that suffering. Nor does it ignore or minimize the moral and political failures that exacerbate it. We will say this over and over, so there can be no misunderstanding: Strategies for a sustainable human presence must involve holding the wealthy and powerful accountable for damage done and moving toward a more equitable distribution of wealth and power. Those goals are desirable independent of ecological realities. Those realities also mean, as a starting point, a commitment to a down-powering and an acceptance of limits, which is necessary for the withering away of the growth economy, which is necessary for long-term survival. Call it "degrowth" or "steady state economics" or "doughnut economics."[7] Advocates for different approaches will disagree about specifics of policy proposals, but there is growing awareness of the need to talk about limits. That starts with recognizing the need to transcend capitalism and the current politics designed to serve capitalists, in pursuit of an equitable distribution of wealth within planetary boundaries.[8] Those of us living in the more affluent sectors of the world should not try to evade these moral assessments and political obligations.

If this kind of honest reckoning with history and contemporary economic-political realignment were accomplished, then what? With nearly eight billion people and most of the world's infrastructure built with, and dependent on, highly dense energy, then what? If running that existing infrastructure on renewable energy is highly unlikely, then what?

It's tempting to believe that we can identify low-energy societies from the past or communities in the contemporary world with lower-energy living arrangements and then simply replicate them more widely. It's tempting to believe that breaking the grip of concentrated wealth and power and expanding democratic decision-making will lead to sustainable societies. But such hopes are based on a misunderstanding of the problem.

Our task today is not only to learn how to live "lower on the food chain," but how to transition from the existing infrastructure and organization of contemporary societies to infrastructure and organization that is consistent with a sustainable future. And we have to do this living with population densities far greater than any previous phase of human history, with an eye toward dramatic reductions in population. No past or existing society or ideology provides a workable model or viable plan for this task.

In those efforts we should learn from the low-energy societies of the past and exemplary experiments within today's societies while we craft a new moral and ideological grounding for a down-powering society. But those good examples don't offer a program for moving from the current state of most societies (large populations, high-energy, unsustainable) to where we need to be (smaller populations, low-energy, sustainable). There are very few examples in history of a complex society choosing to scale back. For example, the eastern portion of the Roman Empire, what came to be known as the Byzantine Empire, survived after the fall of the western portion by choosing to reduce the complexity of their society. But the task before us today is far more daunting: a down-powering on a global level with the goal of fewer people living on less energy, achieved by means of democratically managed planning to minimize suffering. Daunting, indeed. In chapter 2, we discuss those goals and challenges.

No one has yet offered a program to achieve the task before us. Simply invoking previous societies that lived with less energy and lower population densities is not a program. Because planning for transition on this scale is difficult to imagine, people are quick to embrace technological optimism and imagine that we will invent our way to a just and sustainable future without harsh reckoning and dramatic realignment. This optimism slides all too easily into a technological fundamentalism that undermines people's ability to acknowledge and face the difficult challenges. We say more about this fundamentalism in chapter 2, but for now we simply want to highlight the tendency, from people on all sides of the political debate, to seek salvation in a faith-based claim that the development of more high-energy advanced technology will resolve vexing energy and resource challenges. This optimism allows the fantasy that societies can continue at existing high-energy levels through endless innovation that can be fueled by low-cost renewable energy that will become abundant enough to replace fossil fuels.[9] Innovation and renewable energy are important, and we're all for expanding research and development. But we have to do full-cost accounting on renewables and recognize that there are destructive environmental and social consequences to constructing the infrastructure for that energy production.[10] As one researcher puts it, "While the sun and wind are indeed infinitely renewable, the materials needed to convert those resources into electricity—minerals like cobalt, copper, lithium, nickel, and the rare-earth elements, or REEs—are anything but."[11] And independent of the ecological costs of the mining, "global reserves are not large enough to supply enough metals to build the renewable non-fossil fuels industrial system or satisfy long term demand in the current system."[12]

We—that is, all humans—should reject the fundamentalist faith in technological solutions that is all too common on the Right and the Left, as well as the Center. We should reject such delusions, which hinder real progress toward a sustainable future. Rational planning for down-powering that accepts biophysical limits is necessary. We use the term "rational" with "planning" with some hesitation, well aware that much of the planning of the industrial era that was believed to be rational—that is, based on evidence generated while working within widely accepted scientific theories—is the source of many of our most vexing problems. In retrospect, a whole lot of rational planning of the recent past looks distinctly irrational. But we see no option but to embrace a rational process, evaluating evidence without delusions of grandeur and learning from the many mistakes of the past.

Such rational planning means we cannot pretend that if we humans were freed from hierarchical social systems we would suddenly find it easy to avoid the comforts and pleasures associated with dense energy, to which people have become accustomed (in the more affluent societies) or to which they aspire (almost everywhere else). While much wasteful consumption is driven by the propaganda of the growth economy (i.e., advertising and marketing), fossil fuels and other sources of energy also make people's lives easier in many ways that are not frivolous. There is considerable variation in people's assessment of their needs, but capturing and using dense energy for comfort and pleasure is not a unique goal of imperialists and capitalists.

We return to these problems in chapter 2, but for now we suggest that there are no solutions, if by solutions we mean ways to support anything like the existing number of people at anything like the existing level of aggregate consumption. Wishing it to be possible, simply because the alternatives are difficult to imagine—let alone achieve—does not make it possible.

The Need for Justice

We live amid dramatically different levels of energy consumption, resource exploitation, waste production, and overall contribution to ecosystem instability. This highly skewed distribution of wealth is a product of the domination/subordination dynamics that have taken hold over the past ten thousand years, since the invention of agriculture and the emergence a few thousand years later of patriarchy. Most relevant today is the barbarism of European nations in their world conquest over the past five hundred years (typically rationalized with an ideology of white supremacy) and ongoing economic domination in the postcolonial period, for when imperial armies go home, private firms continue to exploit resources and labor, typically with local elites as collaborators.

At the heart of the unsustainable nature of human economic activity is the carbon imperative, the drive to obtain the benefits that come from using dense energy. The dominant vehicle for that destructive extraction today is a rapacious transnational corporate capitalism and that system's requirement of unlimited growth in the pursuit of profit. This is not the secret plan of a greedy cabal but rather the stated goal of the system. When he was deputy treasury secretary in the Clinton administration, Lawrence Summers said, "The potential for the American economy to grow is unbounded. . . . We cannot and will not accept any speed limit on American economic growth. It is the task of economic policy to grow the economy as rapidly, sustainably and inclusively as possible."[13] The possibility that rapid economic growth and sustainability are incompatible is rarely acknowledged. Summers has a particular gift for tone-deaf arrogance, but his comments are unremarkable. His embrace of irrational exuberance didn't get him fired from his deputy position. Instead, he went on to become treasury secretary and president of Harvard University, returning to government service as director of the National Economic Council in the Obama administration.

Because capitalism is, and always has been, a wealth-concentrating system, a relatively small number of people reap most of the financial benefits from the ecological destruction that comes with modern economic growth. In short, the first world is rich, and much of its wealth is concentrated in the hands of a relatively small segment of those societies' populations.

Some people who benefit from these arrangements are dedicated to maintaining the hierarchical systems at the heart of the unsustainable economy and its unjust distribution of wealth. Other people who benefit may condemn those systems but take no action to disrupt them. And some people work for change. We all should do a self-inventory, uncomfortable though it may be, to assess honestly where we fit in these categories. We all should resist the urge to stake out the moral high ground for ourselves, even if it is warranted.

All of the nearly eight billion people in the world are not similarly situated, and this inequality within the human family must never be overlooked in our analysis. But when analyzing the ecological crises facing humans today, we must talk about a "we" that includes everyone—on biological, historical, philosophical, and political grounds.

We Are One Species

It should be uncontroversial to assert the antiracist principle that we are one species, which is anchored in basic biology. There are some observable differences in such things as skin color and hair texture, as well as some patterns in predisposition to disease based on ancestors' geographic origins, but the idea of separate races was created by humans and is not found in nature.[14] There are no known biologically based differences in intellectual, psychological, or moral attributes between human populations from different regions of the world. There is individual variation *within* any human population in a particular place (obviously, individuals in any society differ in a variety of traits). But there are no meaningful biologically based differences *between* populations in the way people are capable of thinking, feeling, or making decisions. As we will continue to emphasize, we are one species. We are all basically the same animal, which most of us understand without reading the scholarly literature. A leading neuroscientist puts it this way:

> I can't say this loudly enough: *There is no evidence for significant average differences in intelligence-related genes between "races." Not between self-identified whites and blacks in the United States, nor between any pair of self-defined racial groups.* Not only that, there is no evidence for racial group differences in genes that have been linked to *any behavioral or cognitive trait.* Not aggression. Not ADHD. Not extraversion. Not depression. Nada, *niente, nichts,* bupkis.[15]

Although we are one species, there are obvious cultural differences among human populations around the world. Those cultural differences aren't a product of human biology; that is, they aren't the product of any one group being significantly different genetically from another, especially in ways that could be labeled cognitively superior or inferior. So why have different cultures developed in different places? The most obvious answer is that it is the result of humans living under different material conditions. Other possible explanations for variations in cultures include a supernatural force providing divine guidance or simple randomness. Theological explanations—that there is some nonmaterial force that dictated or set these patterns in motion—are based in faith claims and don't rely on evidence. We have never identified any compelling reasons to accept supernatural accounts of natural phenomena. Nor have we ever heard a coherent argument for how cultural differences are simply random. So we conclude that the type of living arrangements that groups of humans develop arise from the differences in geography, climate, and environmental conditions. Absent any other credible explanation, we assume that the different material realities under which humans have lived have shaped the variations in human culture. People make choices to build cultures in specific ways, but if all people are basically the same animal, then the differences in those choices around the world are most likely the product of those different conditions.[16]

We know that we make decisions, individually and collectively, in ways we do not and cannot fully understand. Our experience of freely choosing does not mean that all of our choices are 100 percent freely made. Without attempting to resolve the age-old debate on free will, all of us can reflect on how often we come to recognize that past choices, which we believed we made freely at one moment in time, were shaped and constrained by material conditions that we could not understand at that moment and may never fully understand. While we continue to act day to day on the assumption of free will, we also should continue to be alert for ways in which behavior is to some degree determined.

In short, we need to use whatever free will we have to understand the determinism that is at work to shape our choices. This is of course a logical conundrum, but it is still an apt description of the human condition. Centuries of philosophical and scientific inquiry haven't done much to change this. We try to deepen our understanding of deterministic forces while living as if we have expansive free will. That doesn't end the debates about free will and determinism, but it captures our experience.

What are the implications of all this? As we've already said, before we condemn the unsustainable and unjust actions of others, we should be critically self-reflective about our own contributions to the current degraded state of the ecosphere and the inequality around us. That's the first step. The second step is to go beyond the failures of individuals to assess the political and economic systems that reward pathological behavior and impede virtuous behavior, especially the systems we live in and tend to take for granted. The third step is to think historically, recognizing that any group of humans living under the same material conditions would most likely have developed in roughly the same way. There is nothing intrinsically special about any one of us or any one group of people.

This cautious approach is a way of extending the adage "There but for the grace of God go I" beyond individuals to cultures. That phrase emerged from a Christian assertion of humility in the face of God's mercy, but we use it here in a secular fashion. If one has lived an exemplary life, that's great, but be aware that life might have been very different if some of the material conditions in which one lived were different. Those who believe they have accomplished something and made a positive contribution to the world should remember that a change in any one of the conditions in our lives, especially in our formative years, may have meant failing instead of succeeding. We are not suggesting that we have no control over our lives but simply that we likely don't have as much control as many people would like to believe.

This is true of us individually and collectively. The conditions under which a culture emerged may have led to ecologically sustainable living arrangements, but those living arrangements would have been different if initial conditions had been different. If Culture A created an ecologically sustainable way to live and Culture B created an unsustainable system, it is important to highlight the differences, endorse Culture A, and try to change Culture B. But if the geography, climate, and environmental conditions out of which the two cultures emerged had been different, then what would A and B look like?

In our secular analysis, there but for the specific geography, climate, and environmental conditions go *we*. For example, because of the differences in initial conditions, not all cultures developed the technologies to plow the ground, smelt ores, or exploit fossil fuels to do work in machines. The cultures without those technologies have not depleted the carbon in soils, forests, coal, oil, and natural gas in the ways that societies with those technologies have done. The development of those technologies was not the product of inherently superior intelligence of people in particular regions of the world. Remember, we are committed to an antiracist principle that flows from basic biology. That means the forces that led to the creation of those technologies must have been generated by the specific environmental conditions under which that culture developed over time. Likewise, the lower rate of carbon depletion that results from the absence of those technologies cannot be a marker of inherently superior intelligence of people in particular regions but is instead the product of environmental conditions. In a significant sense, the trajectory of people and their cultures is the product of the continent and specific region in which they have lived.

Many who consider themselves antiracist might bristle at this analysis. So we want to be clear about how we understand racial and ethnic differences in the context of political and economic history. Europe is not rich because Europeans are racially superior. Europe is rich because it developed on a different trajectory from that of the Americas,

Africa, and Asia as a result of geographic and environmental differences. That trajectory made it possible for Europeans to conquer and exploit the people and resources of those other continents. At one point, Europeans believed themselves intellectually and morally superior because of racial differences that were assumed to be immutable. We know that to be false. But if that's false, then so is any other claim by any other group to be intellectually or morally superior on any criteria by virtue of a racial or ethnic identity.

If history was not shaped by the minor genetic differences that are associated with our ancestors' region of the world, that leaves us with geography, climate, and environmental conditions, unless we want to argue that history is directed by God or gods or is simply random. We really are one species.

Scholars who have presented compelling data and arguments for what is typically called geographic or environmental determinism point out that these forces do not act in simple, linear fashion.[17] Geography shapes people, and people act to shape the meaning of geography, making choices along the way. But not all people throughout history and around the world have been presented with the same choices by the landscapes on which they have lived. Again, we need not resolve the larger philosophical debate on free will versus determinism to recognize that these material realities are a driving force in shaping human history.

This shouldn't be a surprising claim. All organisms adapt to, and are shaped by, their places. There is no reason that humans should be exempt from that observation. While it's true that humans' physiology and cognitive capacity allow us to live almost anywhere on land on Earth, that doesn't mean that geography has no relevance in how we have organized societies and developed new technologies.

The authors have met many people—including those who share our point of view on social justice and ecological sustainability—who are nervous about any exploration of this analysis. This resistance seems based in the fear that acknowledging the role of geography in human history somehow denies people any sense of agency and/or provides absolution to those people who have exploited other people and the nonhuman world. We understand that fear but believe that understanding contemporary problems and planning for the future requires that we not ignore relevant information and analyses. Below is one suggestion for overcoming those fears.

Thought Experiment I

We cannot do controlled laboratory experiments on history, of course. But thought experiments—the "what if" explorations—can be instructive, not to prove a point in a scientific sense, but to help us think more clearly and expansively.

We have highlighted how some Europeans over the past five centuries have exploited people and landscapes to enrich themselves, to the detriment of people, other organisms, and ecosystems. One of the primary sites of exploitation was the Americas. Today, in the United States, people rightly point to the more sustainable lifeways and economies of the varied Indigenous North American cultures compared to the European record of human exploitation and ecological degradation. There is, of course, considerable variety in Indigenous North American societies, a product of the varied land and water resources on which those societies developed. We agree with the common warning against romanticizing those societies, but it is accurate to emphasize that the living arrangements developed in North America were more sustainable than those developed in Europe in the modern period. Such comparisons are important, and as we have argued, the dominant culture should learn from any society that has lived successfully at lower levels of energy and resource consumption, recognizing that those lifeways developed with much lower population densities than those that exist today.

But what if Indigenous North Americans had lived with the climate and geography of Europe and vice versa? Imagine going back ten thousand years and flipping the geography, climate, and environmental conditions of western Europe and the continental United States. The people living in both places stay the same, but the local environments are swapped and the conditions in the Atlantic Ocean reversed. What would the world have looked like in 1492? Would the state of societies in those places have been the same in 1492? What would the world look like today?

The answers seem self-evident. If the environmental conditions had been flipped—assuming an antiracist position that understands there are no meaningful differences between population groups—we would expect the developmental trajectory of the societies would have been flipped. Europeans would not have become the imperial master. North Americans would have been in a position to dominate the globe because of their ability to capture more energy per person; to develop more advanced technologies, including the tools of war; and to move around the globe. Would North Americans who had the opportunity to become the imperial master have rejected that option? Remember that in this thought experiment they would have developed in the same geography as Europeans actually did develop. What reason would we have to assume their development would have been different? If they had lived in the geography of Europeans, would they not have developed in pretty much the same way Europeans did and vice versa?

As we have pointed out, some might invoke a theological explanation, that a particular belief system provided guidance that led to a different path. But unless one endorses a supernatural explanation—and we are committed to a secular worldview that does not look to causes beyond the natural ones—this spiritual explanation also leads us back to differences in geography, climate, and environmental conditions. If the differences in cultural stories about the spiritual realm don't come from a supernatural source, they are either random or the product of people developing in specific places. We assume the latter.

Our point is not that individual Europeans had no choices over the past five hundred years. The fact that not every person in Europe endorsed the path of conquest reminds us of the individual variation within all societies and the complexity of history. Again, we aren't trying to resolve the unresolvable debate over individual free will but are simply pointing out the power of geography, climate, and environmental conditions in determining the direction of human societies.

We also are not saying that the descendants of those earlier Europeans—people like Jackson and Jensen—are off the hook today for the disparities in wealth and power. While we are not responsible for the actions of people who lived before we were born, we are responsible for our actions today, in a world shaped by that history.

Now we turn to a crucial step in this analysis. Any recognition of the role of geography in shaping the development of different human societies raises an obvious question: Does geography still dictate our fates today? If we accept a geographic-environmental determinist argument, does that mean there is no point in trying to create sustainable societies? Are we at the mercy of forces completely beyond our control? Our answer is a cautious "No."

We believe that increased knowledge and critical self-reflection about these matters increases the scope of human agency. As a species, we have learned much about the consequences of past action, which creates the opportunity to change course. But to do that, we must understand what shaped our past. Accepting that there is some level of determinism in our lives reminds us that if we are to achieve sustainability, we have to build social and ecological change into our lives at a deep level. Rejecting the most destructive economic systems, such as capitalism, is a necessary first step but not sufficient to deal with the problem. We have to create new systems that take seriously the temptations of dense energy. Those temptations can—and often have—overwhelmed people's capacity for rational planning and moral decision-making.

Our thesis, restated, is that all people are members of the same species, and just as with any organism, we have species propensities that shape our activities. Like all animals, we seek energy-rich carbon. As an animal with considerable cognitive capacity and the ability to cooperate in large-scale and complex ways, we've gotten exceptionally good at getting lots of that carbon. Human carbon seeking has varied depending on the kind of places in which we have lived, but since the invention of agriculture that quest for carbon has often been more destructive than anyone could have predicted at the outset.

Human Nature and Humility

Recognizing that material realities shape our lives does not absolve anyone or any society today of moral accountability for their actions or for the failure to act. Recognizing that some level of environmental determinism shapes history is not a free pass for those of us who hold a disproportionate share of the world's wealth and who are responsible for a disproportionate share of ecological destruction. Whatever people knew about the ecological consequences of their actions when contemporary cultures first developed, we now know more than enough to act on what our own moral principles demand of us—pursuing living arrangements consistent with social justice and ecological sustainability.

The first farmers who used plows could not have understood how devastating soil erosion would be. The first metalworkers could not have understood the consequences of the destruction of forests to smelt ore that helped create and maintain empires. The first inventors who figured out how to burn coal in steam engines to replace human and animal muscle power could not have predicted global warming. But today we do know, and if we fail to live up to the principles we claim to hold, then we appropriately are the targets of demands for corrective action.

But in trying collectively to find a way out of the mess we've made, the assigning of different levels of responsibility for the mess is only a first step. No individual, political movement, or government has a viable plan for transitioning from an unsustainable high-energy, interdependent global society of nearly eight billion people to low-energy societies with sustainable levels of population and consumption. While lessons from low-energy societies will undoubtedly be valuable, there is no way to flip a switch and return to a previous era's living arrangements and lower population densities. Technological innovation and renewable energy will play a role but cannot power the infrastructure of a world built with the highly dense carbon of fossil fuels.[18]

And while our economic and political systems must change, we have to acknowledge that breaking the grip of concentrated wealth on politics won't change the fact that dense energy makes our lives easier in many ways that most people enjoy and will not want to give up. Just as geography is relevant in understanding the path of history over long periods, human nature is relevant in understanding ourselves today. Unless we believe humans are different in kind from all other animals, we have species propensities that must be considered in planning and setting policy. It should not be controversial to state that the possible range of behaviors for all animals is determined by their genetic code. One of those behaviors shared by all organisms is the tendency to maximize the energy they can capture. The ability to transform energy into work, to maximize the flow of energy, determines fitness and success in evolutionary terms. Evolution looks kindly on organisms that maximize their use of power.[19]

Just as people advocating for social justice and ecological sustainability typically are wary of geographic and environmental determinism, they are often nervous about talk of human nature. That's at least in part due to capitalists' success in narrowly defining our nature as inherently greedy and self-interested. Humans have the capacity to act in greedy and self-interested fashion, of course, but capitalism's critics point out that we also have the capacity to collaborate and cooperate. That's not only part of our nature; it has been a crucial factor in human expansion across the globe.[20]

It is ironic that the philosopher most associated with capitalism's origin story, Adam Smith, rejected the crude version of human nature that later capitalists would trumpet. Seventeen years before *An Inquiry into the Nature and Causes of the Wealth of Nations*, Smith published *The Theory of Moral Sentiments*, which includes astute observations about the complexity of human nature.[21] Smith writes:

> How selfish soever man may be supposed, there are evidently some principles in his nature, which interest him in the fortune of others, and render their happiness necessary to him, though he derives nothing from it except the pleasure of seeing it. Of this kind is pity or compassion, the emotion which we feel for the misery of others, when we either see it, or are made to conceive it in a very lively manner. That we often derive sorrow from the sorrow of others, is a matter of fact too obvious to require any instances to prove it; for this sentiment, like all the other original passions of human nature, is by no means confined to the virtuous and humane, though they perhaps may feel it with the most exquisite sensibility. The greatest ruffian, the most hardened violator of the laws of society, is not altogether without it.[22]

Without this species propensity for collaboration/cooperation and sympathy/empathy, we wouldn't be writing this book because no one would be writing books because humans would not have come to be the dominant species on the planet. Peter Kropotkin—an anarchist and a scientist[23]—used the term "mutual aid" to capture the power of reciprocity and cooperation that is mutually beneficial.[24] But like Smith, he did not deny the other aspect of our nature, what he called a "double tendency" within human beings, and coping with that was at the core of his ethics.

> The chief demand which is now addressed to ethics is to do its best to find through the philosophical study of the subject the common element in the two sets of diametrically opposed feelings which exist in man, and thus to help mankind find a synthesis, and not a compromise between the two. In one set are the feelings which induce man to subdue other men in order to utilize them for his individual ends, while those in the other set induce human beings to unite for attaining common ends by common effort: the first answering to that fundamental need of human nature—struggle, and the second representing another equally fundamental tendency—the desire of unity and mutual sympathy. These two sets of feelings must, of course, struggle between themselves, but it is absolutely essential to discover their synthesis whatever form it takes.[25]

None of this requires us to be nostalgic or naive. In the words of a contemporary philosopher with an eye on human evolution:

> Egalitarian, cooperative human communities are possible. Widespread sharing and consensus decision-making aren't contrary to "human nature" (whatever that is). Indeed, for most of human history we lived in such societies. But such societies are not inherently stable. These social practices depend on active defence. That active defence failed, given the social technologies available, as societies increased in scale and economic complexity.[26]

Capitalism has done its best to ignore the tendency toward sociality and mutual aid, in the service of an ideology that asserts that everyone is better off if people act only on a narrowly defined material self-interest.

But the fact that some people offer a selective account of human nature to justify wealth-concentrating economic policies does not mean that a more honest and expansive reflection on human nature should be ignored in formulating policy.

It should not be necessary, perhaps, to defend the idea that human nature exists and is relevant to our inquiry, but because of this history of distortion or avoidance of the subject, we will state the obvious: there exists something we can call human nature, just as there is hummingbird nature and wolf nature and chimpanzee nature. That simply means that every organism has a genetic endowment that makes some things possible and some things impossible. There are parameters within which all organisms, including we humans, operate. That said, everyday experience reminds us that human nature produces widely variable behavior and that there is little we can predict with certainty about any specific human's behavior in a particular situation.

Take the example of whether violence is a part of human nature. There is no reason to believe that any human society has been completely aggression-free. We are a species capable of violence toward one another. It's likely that all humans—even those who may never have been violent in their lives, if such a human has ever existed—have the capacity for violence against others. Under what social conditions is violence more or less likely? What individual differences, interacting with those social conditions, might increase or decrease the likelihood of violent action? Socialization shapes the expression of human variation, and there are patterns in how people respond to that socialization. We never know as much as we would like to know about these kinds of questions and are usually left to act on informed hunches based on limited evidence.

The perception of patterns in the complex way human nature plays out is the best we can hope for when trying to understand ourselves, our behavior, and the social norms that shape that behavior. We try to understand the parameters set by biology and do our best to discern the patterns within those parameters. Since anything that human beings do is, by definition, within our nature to do, the question, "What is human nature?," should be replaced with questions about which aspects of our nature tend to dominate under various conditions. We agree with the economist Elinor Ostrom who, in her 2009 lecture accepting the Nobel Prize, observed that "humans have a more complex motivational structure and more capability to solve social dilemmas" than mainstream economists recognize and that "a core goal of public policy should be to facilitate the development of institutions that bring out the best in humans."[27] But to bring out the best in us requires us to know what kind of creatures we are, not to indulge the fantasy that, unlike all other organisms, humans are some kind of blank slate.

Our focus here is on a particular aspect of human nature, what we might call our human-carbon nature, a phrase borrowed from our colleague Bill Vitek.[28] This reminds us that we are carbon based like all other life on Earth. What is life? What is the nature of living things? Scientists list various characteristics that mark living things, such as the capacity to grow, metabolize, regulate the internal environment, respond to stimuli, adapt to the environment, reproduce, and evolve. Our answer for purposes of this inquiry is a definition that Jackson has been offering for some time: "Life is the scramble for energy-rich carbon."[29]

It is our human nature, like the nature of all life, to seek out energy-rich carbon. To be alive is to go after carbon. Over time, humans have gotten exceedingly good at tapping into five major carbon pools—soils, forests, coal, oil, natural gas—and maximizing the extraction of all the carbon we can get our hands on.[30] There are few exceptions to that pattern. Our greatest success as a species has become our most profound failure, given the many negative consequences of all that carbon grabbing. Understanding and changing our response to our human-carbon nature is a new challenge for which we have no road map. No existing ideology or culture is going to provide us with a template for dealing with what lies ahead.

Renouncing first world dominance is a start, as is imagining a world beyond capitalism's obsession with growth and consumption. The end of those systems is a necessary but not sufficient condition for change. We've stated our agreement with the ecosocialist goal "system change not climate change," but we disagree with that network's claim that "the current ecological crisis results from the capitalist system, which values profits for a global ruling elite over people and the planet."[31] The current ecological crisis is shaped by, but not the result of, capitalism. Human degradation of ecosystems predates capitalism and will continue after capitalism, unless we develop a deeper understanding of the crisis.

As one scholar of collapse puts it, "History suggests that complexity most commonly increases to solve problems, and compels increase in resource use."[32] The end of capitalism won't necessarily disrupt that pattern. Placing our hopes in noncapitalist complexity will not magically reduce resource use. So it's not enough to reject capitalism. We have to confront our own carbon-seeking nature. If we start with an awareness of the scope of the change needed and the lack of a plan for dealing with human-carbon nature, we can at least be clear about the direction in which we need to move. And that requires committing to being the first species that will have to impose limits on itself, which means a collectively imposed cap on the carbon we use and rationing to ensure fairness.[33]

To put it as bluntly as possible: Any policy that does not understand and account for the temptations of dense energy will fail. Human-carbon nature matters.

Thought Experiment II

Another "what if" thought experiment is instructive here, to challenge those in the social justice and sustainability movements who not only are nervous about acknowledging the geographic determinism in history but also are resistant to bringing human-carbon nature into policy debates. We routinely talk with people who assert that the problem is not that there's too much aggregate consumption but that the distribution of that consumption across the human population is unequal and inequitable and that capitalism creates wasteful consumption by manipulating human desires in pursuit of profit. Those observations about capitalism's unjust and wasteful character are accurate, but they don't undermine the importance of asking critical questions about consumption more generally.

To be clear: We are anticapitalist, on moral, political, and ecological grounds. Capitalism, with its growth imperative and wealth concentration, has proved to be inconsistent with basic human decency, democracy, and sustainability. But this assessment shouldn't lead to a demand for political purity today. Given the global dominance of capitalism's regime of ownership and finance, for now, any strategy for advancing justice and sustainability has to account for that power and maneuver within that system. We have no plan for vanquishing capitalism and are open to any and all creative proposals for change. Still, we believe it's crucial to point out the pathological nature of capitalism and endless-growth economics, both to guide immediate action and to keep us focused on the eventual end of that system.

As a first step, we support egalitarian principles that are central to socialism. But our vision of a just and sustainable future includes a rejection not just of the capitalist worldview but also of the industrial worldview's expectations for expansive energy consumption. We do not think that even a well-designed socialist system is up to the challenge in front of us, unless it emphasizes the need for collective self-imposed limits on human energy expectations. Why? Because of our human-carbon nature.

Let's go back to when humans first discovered how to use fossil fuels to do work, the time of the first steam engines that burned coal to replace the muscle power of people or animals. Prior to that, in some locations water power had replaced some of that labor, but steam engines and coal ratcheted everything up several notches. Imagine that in that moment, the world had been structured on socialist principles in a socialist economy with meaningful democratic decision-making instead of an incipient capitalist economy and elite dominance. Would socialist societies have rejected the use of coal to do work? Would socialists have said, "Wait, the unprecedented energy in this highly dense carbon is going to dramatically change human relationships to the natural world, our relationships to each other, and our sense of our own needs"? Would socialists have concluded that the potential negative consequences were not worth the risk? Is there any reason to believe that socialists would not have acted from the species propensity to maximize the amount of carbon we could extract from the environment?

Allow us a short digression. Some might cite the Luddites of early nineteenth-century Great Britain as a counter to this claim. This loose group of artisans and craftspeople rebelled against the imposition of a new economic and technological system taking shape in factories that undermined the viability of their largely home-based workshop production. At the heart of the machine breaking of the Luddite rebellion was not a rejection of technology using more energy but a fear of the way in which technology yoked to industrial capitalism would, and did, impoverish their lives. They were not against machines per se but against "all Machinery hurtful to Commonality."[34] The Luddite resistance should be honored for an early challenge to industrial capitalism but is not evidence that socialism would have rejected high-energy technologies.

We have no doubt, however, that a well-designed socialist society would have used that energy for different, and more socially beneficial, purposes than a capitalist society. Instead of maximizing return on capitalists' investment, a socialist system could seek to maximize human flourishing for everyone. The dense energy of fossil fuels could have been put to better use. Instead of luxury consumption for some, the goal could have been more material comfort for all. But there's no reason to believe that socialist societies would have resisted all the temptations of all that dense energy. Nor is there reason to believe that a more egalitarian system today would be able to limit ecological destruction in significant ways, unless it embraced a collective rejection of the contemporary high-energy "lifestyle" and prioritized a collectively imposed cap on the amount of carbon we use. Yet this component of a viable plan for ecological sustainability—a clear statement of the need to dramatically reduce human aggregate consumption—is either absent or downplayed in current socialist and ecosocialist programs. Instead, these programs tend to suggest that continued development of renewable energy will solve our problems without a dramatic reduction of economic activity.[35]

In contemporary society we are surrounded by uses of energy that are, by anyone's standards, wasteful and often grotesque. A handful of ultra-rich folks partying on a yacht is wasteful, and when the partying goes on while so many other people suffer it is grotesque. But the fossil fuels that make that yacht possible also can be used to do work that lessens the physical demands on people.

Think of the work of moving water from a well to a dwelling that is one hundred yards away (about the distance from the well to the house in which Jensen lives today). Digging the well by hand, carrying the water by hand—all of that takes a lot of human labor. In the industrial world, that well can be dug with machines, which also can dig a trench to lay a pipe that will carry that water to the dwelling. Would it be easy to reject the help of fossil fuels to do those jobs? Even if we were willing to forgo the backhoe and pick up a shovel, who wouldn't prefer a steel-bladed shovel over a stone digging tool? Would living in a socialist society make it somehow easier to reject the products of the industrial world and do it all by hand? A preference for the industrial solutions made possible by the dense energy of fossil fuels is not the product of capitalist indoctrination. It's just easier on one's back.

High-energy tools that make our lives easier come with a cost, and we should always evaluate not only the benefits, but those costs. Everyone would agree with that in theory, but people routinely avoid those calculations. It is especially difficult to give up labor-saving tools once we have them, which suggests that we should think carefully before introducing new high-energy tools. But that kind of critical evaluation of new gadgets is rare.[36] Given the current crises, we should constantly look for places to abandon high-energy tools in favor of lower-energy methods and reassess the need for the work those high-energy tools do. In the contemporary United States, we have yet to see such questions asked. Critics of capitalism are quick to challenge gratuitous consumption, the kind of spending that does little but generate corporate profits. That's the low-hanging fruit. More important is grappling with the hard questions about limits and rationing.

Transcending capitalism is a necessary but not sufficient condition for achieving just and sustainable human communities. The concentrated power in a capitalist system is unlikely to shape sustainable societies. The more democratic decision-making possible in a well-designed socialist system offers a path for rational planning. But it is naive to believe that such a system will make it easy to impose limits, especially when the techno-optimists tell us that we can have it all.

We return over and over to an insightful observation that captures this problem, borrowing from George Orwell's critique of left-wing parties in Britain. Orwell pointed out that those parties had internationalist aims but also wanted to maintain a standard of living that depended on exploitation around the world. Similarly, we recognize that environmentalists—including us—are hooked on dense energy just like almost everyone in contemporary society and therefore "make it their business to fight against something which they do not really wish to destroy."[37]

Animals All

Critical scholars have helped us get beyond the "great man" theory of history, the idea that history is made by exceptional leaders. Feminist critics have helped us get beyond the "man" theory of history by exposing the patriarchal domination at the heart of civilizations. Social history has moved toward embracing the story of all people and not just elites, what is sometimes called history from the bottom up. That's all to the good, but one more step is required: a history that accounts for people as animals, a deeper analysis of how—like all other organisms—humans adapt to different ecosystems and develop behaviors shaped by geography, climate, and environmental conditions, along with our human-carbon nature.

Some fear that this view of history erases human agency and reduces history to biology. Indeed, the misuse of biology in the past—think of theories of "scientific" racism from the previous century, for example, and the vestiges that still exist—is a good reason to be cautious. While philosophers continue to debate whether we have free will, we live and ponder our options and choose, all based on the assumption that those choices are real and that they matter.

But consider this: whatever one's philosophical view or political position, it is typical that the further back in time we look at human behavior, the more we talk freely about the forces that shape history independent of individuals' choices and intentions. Discussions of the development of human societies and practices in prehistory take as a given the role of evolution by natural selection and environmental forces. For example, in the debate over the role of humans in killing off megafauna,[38] scholars assess the evidence concerning human hunting practices that may have driven, or contributed to driving, those species to extinction. But you don't hear anyone criticizing those long-ago humans for making immoral choices. Instead, we discuss those choices as responses to environmental conditions shaped by human nature. Well, those humans were indistinguishable from us. We're the same animal. But if a group of people today were to hunt a species to extinction, we would not hesitate to make moral judgments. The closer we get to our moment in history, the more we emphasize the choices and intentions of individuals, recognizing that those choices are structured by cultural and ideological systems, which are themselves assumed to be the product of human choices. We should acknowledge that there is a certain amount of vanity in this tendency to exempt ourselves from determinism while accepting it in past cases.

We are not arguing that we should suspend such judgments today but simply want to make sure that we consider the role of environmental conditions and species propensities in assessing human behavior. Social critics rightly point out that a focus only on individuals' behavior obscures the role of social systems in shaping human behavior. In the same way, a focus only on social systems obscures the role of our species propensities. We experience ourselves as making choices about how to obtain energy from the ecosystems in which we live, and that subjective experience of choosing cannot be ignored. But our human-carbon nature makes some choices more inviting than others, and we humans tend to make choices that deliver the most energy at the least cost given existing technologies. That's not a "law" of nature—not everyone makes the same decisions in similar circumstances—but is rather an identification of the patterns in human history.

By recognizing the role of geography and organisms' scramble for energy-rich carbon, we will better understand human history and increase the chances that we can make decisions today that make possible a better future. What is "better" in our view? Social justice and ecological sustainability, people living in egalitarian communities that accept limits on population and consumption, in societies that have learned from the study of history and biology.

When we integrate history and biology, we see the importance of confronting failures today while also taking the long view. That means rejecting first world dominance and capitalism but also going deeper. The problems we face aren't simply the consequence of the past 250 years of fossil-fuel use or the past 500 years of European colonialism. We have to go back further, to the origins of agriculture 10,000 years ago, the crucial fault line in human history. That's when humans began exploiting carbon beyond replacement levels, particularly where grains such as wheat and barley were farmed by annual plowing that disrupted and degraded the soil.[39]

Agriculture did not develop in the same way everywhere on the planet. Variations in geography, climate, and environmental conditions meant that people farmed in different ways in different places, of course. But especially where cereal grain crops dominated, the ability to create and store surpluses generated the expansive hierarchies that have produced profound social inequality. Because grain ripens above ground and must be harvested at a specified time of year, it was an easy target for taxation and regulation of early states.[40] Grain stores required the protection of city walls and armies and could feed the people who were building the walls and taking up arms.

Surplus-and-hierarchy predate agriculture in a few resource-rich places,[41] which produced what anthropologists sometimes call complex hunter-gatherers or affluent foragers. Those societies could support larger and denser populations, sometimes with relatively permanent settlements and more entrenched inequality when compared to what are called generalized hunter-gatherers or traditional foragers. But agriculture is the primary fault line in human history because "there's no surplus without storage, and only a limited surplus without farming."[42] It was the domestication of plants and animals that triggered the spread of hierarchy and a domination/subordination dynamic across the globe, and agriculture also was the beginning of the idea that we humans, rather than the ecospheric forces, control the world.

In short, human nature is variable and plastic. When living under conditions that generate surpluses over which people might struggle for control, it's within our nature to abandon the egalitarian features of our gathering-and-hunting history and create hierarchies. It's all part of human nature, all connected to the scramble for energy-rich carbon that is at the center of life on this planet. That is who we are.

The "we" is us, *Homo sapiens,* the primate with the big brain. In assessing human history, when are individual choices the central part of the story? Again, it's worth remembering that no one talks about the individual choices that foragers made a hundred thousand years ago, or fifty thousand years ago. What should we say about the first farmers, the first smelters of ore, the first people who tapped fossil fuels to do work in machines? All of them contributed to the mess we are in but without knowledge of the consequences of their actions. Perhaps we can say of those early carbon seekers, Forgive them, for they know not what they did (Luke 23:34 NIV).

Today, we know considerably more about the consequences of maximizing our extraction of energy and resources from the world's ecosystems, and we need to hold each other accountable for specific failures. But we repeat our caution: the moral high ground is a dangerous place to stand, even when it's warranted. A sense of moral superiority, even when justified by one's choices made today, becomes harder to maintain when one steps back and combines a deeper sense of history with an awareness of evolutionary biology.

TWO

FOUR HARD QUESTIONS

Size, Scale, Scope, Speed

People are sometimes reluctant to ask questions when they suspect that they won't like the answers. The authors have personal experience with this hesitancy to face reality, more than once and at various points in our lives. We assume that everyone reading this can say the same.

How many churchgoers who have doubts about their congregation's doctrine decide to squelch their questions out of fear of losing friends and community? How often do people in intimate relationships avoid confronting tension because they know a problem can't be resolved and speaking of it will bring the end of the relationship? How many people have delayed a trip to the doctor because they know that an examination will lead to a diagnosis they don't want to deal with?

Here's an exercise for all of us. For one day, pay attention to all the forms of denial you practice and see others practicing. How many times do we turn away from reality because it's too hard in that moment to face it? Dare we list the things that scare us into silence? We often avoid hard questions precisely because they are hard.

What we experience individually is also true of the larger culture. There are hard questions that, collectively, we have so far turned away from, either because we have no answers or because we won't like the answers waiting for us. As we have already said, contemporary societies face problems for which there likely are no solutions if we are only willing to consider solutions that promise no dramatic disruption in our current living arrangements. Hard questions often demand that we acknowledge the need for hard-to-accept change.

Our ecological crises cannot be waved away with the cliché that necessity is the mother of invention, implying that human intelligence, perhaps in combination with market incentives, will produce magical solutions. We believe that the most productive way to face today's hardest questions is to focus not only on human creativity but also on human limitations. The techno-optimists emphasize the former, betting we can do anything we set our minds to. Those who lean toward nihilism focus on the latter, suggesting there's no way off the path to ruin. We believe that responsible planning requires careful consideration of both our species potential and our species propensities—not only what can get us out of trouble but what got us into trouble in the first place. Here are four of those hard questions that are essential to confront now:

- What is the sustainable **size** of the human population?
- What is the appropriate **scale** of a human community?
- What is the **scope** of human competence to manage our interventions into the larger living world?
- At what **speed** must we move toward different living arrangements if we are to avoid catastrophic consequences?

Climbing out of the Overton Window

When we have raised these issues in conversation, the most common response is that those hard questions may be interesting, but they have no bearing on what is possible today in real-world struggles for justice and sustainability. The implication is that such questions either somehow don't really matter or are too dangerous to ask. We've heard this not just from people within the conventional political arena but also from environmentalists and activists on the Left. Their argument generally goes something like this:

Those questions raise issues that most people simply will not engage and suggest a need for changes that most people simply will not make. Sensible environmentalists and activists know that you cannot expect people to think about such huge questions when they face everyday problems of living and making a living, which take up most of their time and energy. And what's the point of thinking about these things anyway, when we all know that politicians can only move so far, so fast in our political system? Why ask questions and offer policies that are certain to be ignored?

Sensible people, we have been told, are those who accept the "Overton Window." Named after the late Joseph P. Overton from the Mackinac Center for Public Policy, the idea is that politicians "generally only pursue policies that are widely accepted throughout society as legitimate policy options. These policies lie inside the Overton Window. Other policy ideas exist, but politicians risk losing popular support if they champion these ideas. These policies lie outside the Overton Window."[1]

That can be a useful concept for thinking about what laws might be passed today, but it becomes an impediment to critical thinking when people use it to avoid hard but necessary questions that can't be put off forever. When confronting questions of size, scale, scope, and speed, we encourage people to climb out of the Overton Window to get a wider view of the world, to think not about how human political processes limit what actions are possible today (which they do) but about what the larger living world's forces demand of us (which dictate the material conditions in which we live our lives). When attempting to come to terms with biophysical realities, refusing to look beyond the Overton Window guarantees collective failure. That window certainly exists in the realm of environmental policy: politicians fear the loss of support if they move too far, too fast. But that doesn't exempt anyone from asking those hard questions. The environmental policies that are possible today are important, but we also must recognize that we likely face a dramatically different set of choices in a far more challenging tomorrow. And that tomorrow isn't as far away as we may want to believe.

We realize that asking those four hard questions in the mainstream political arena today is nearly impossible and that the key actors in our current political system will not engage those questions anytime soon. But to cite those two impediments as a reason not to *ever* grapple with those questions in *any* context is not sensible. It's an indication of moral and intellectual weakness. The nineteenth-century Austrian writer Marie von Ebner-Eschenbach put it succinctly: "There are instances in which to be reasonable is to be cowardly."[2]

We believe there are more and more people willing to climb out of the Overton Window. We constantly meet people who are tired of being told they must be "sensible." If we can refuse to be limited by other people's fears—if we can see beyond both a naive techno-optimism and a corrosive nihilism—we create space for a conversation about those questions without having to pretend that we have all the answers. We can make realistic assessments, drawing on science and on human history. But we have to be willing to drop sunny-side-of-the-street fantasies captured in phrases such as "the impossible will take a little while" and "necessity is the mother of invention" while at the same time refusing to slip into a paralyzing despair.

The questions posed above are so complex that detailed answers are beyond our capacities, but that doesn't render them irrelevant. With these caveats, we are confident in asserting the following rough conclusions.

- Size: Earth's ecosystems can sustainably support far fewer than eight billion people, even if everyone were consuming far less energy and material than today.
- Scale: We will have to learn to live in smaller and more flexible political and social units than today's nation-states and cities.
- Scope: We are far less capable of controlling modern technology than we think and cannot manage the current high-energy/high-technology infrastructure we have created for much longer.
- Speed: We must move faster than we have been and faster than it appears we are capable of.

Size

For any species, an ecosystem has a carrying capacity—the population size and density that can be sustained in that place given available resources. No species can increase its population in an ecosystem indefinitely. Eventually, an expanding species will run out of adequate space for its activities; exhaust necessary resources, especially food and water; and be unable to safely dispose of waste products. Eventually, the population of a species that expands beyond carrying capacity will be reduced by starvation, predation, and disease. Every ecosystem has a carrying capacity, which means the planet—a collection of ecosystems—has a carrying capacity for any given species.

No one disputes those statements when applied to nonhuman species. But lots of people believe those rules somehow don't apply to the human species. Because the human population has expanded so dramatically in our lifetimes, it's tempting to believe that we are the only species not subject to biophysical limits. We don't want to caricature those who disagree with us, but that argument usually goes like this:

> A couple of centuries ago, Thomas Malthus predicted that human population growth would undercut any gains in rising standards of living because population growth would outrun food production. Yet the human population has kept increasing, from just under one billion in his time to nearly eight billion today, made possible by dramatic increases in food production. So Malthus was wrong and anyone who says anything like that is also wrong.

As one science writer put it, people who take Malthus seriously "cannot let go of the simple but clearly wrong idea that human beings are no different than a herd of deer when it comes to reproduction."[3]

In the debate between what are sometimes called the neo-Malthusians and the Cornucopians, it's easy to reject the extreme versions of either approach. Some neo-Malthusians have issued specific predictions of coming catastrophes that have proved to be wrong.

The most well-known example is Paul Ehrlich's 1968 best-seller, *The Population Bomb*, which began with this proclamation: "The battle to feed all of humanity is over. In the 1970s and 1980s hundreds of millions of people will starve to death in spite of any crash programs embarked upon now."[4] Some Cornucopians suggest that there is no limit at all to resource extraction and therefore no limit to population growth. Julian Simon's 1981 book, *The Ultimate Resource*, updated and reissued in 1996, includes a chapter titled "Can the Supply of Natural Resources—Especially Energy—Really Be Infinite? Yes!"[5]

We should be wary of predictions, especially about complex processes that are well beyond our capacity to understand fully, and of course a lot has changed since Malthus. The past two centuries have seen increased yields through the use of fossil energy and industrial methods in agriculture, advances in public health, and unprecedented medical innovation. But along with those changes have come rapid climate destabilization and a long list of other ecological crises. The dense energy in fossil fuels and the advanced technology developed since the Scientific Revolution have dramatically increased food production and also dramatically undermined the stability of virtually all of Earth's ecosystems through increased soil erosion and degradation, widespread chemical contamination, species extinction, and climate disruption.

All life—including human life, no matter how technologically sophisticated we become—depends on those ecosystems. What is the sustainable carrying capacity of the planet for *Homo sapiens*? Estimates vary from 500 million to more than one trillion, with the majority of studies putting the number at or below eight billion people.[6] Whatever the answer, the sustainable carrying capacity of the planet for humans has limits, it's likely that humans passed a sustainable limit decades ago, and we are able to avoid the consequences of that reality *temporarily* through exploitation of more dense energy with more advanced technology.[7] We see no reason to believe that can go on forever. This conclusion does not mean we can do nothing but wring our hands while preaching doom and gloom. We can advocate for collectively setting limits as an important part of rational and responsible planning, which has to be based on an honest assessment of the conditions under which we live today and can expect to live tomorrow.

While climate change is an existential threat, a focus only on climate can be misleading. Climate disruption is a derivative of overshoot,[8] of too many people consuming too much stuff in the aggregate. If a miracle solution to climate destabilization appeared tomorrow, we would still face multiple cascading crises because human demands on Earth's ecosystems are in excess of those ecosystems' capacity to regenerate in a time frame relevant to us. Fertility rates are declining in most places, but a slower rate of growth doesn't solve our problems. At some point in the future, there will have to be significantly fewer people and a lot less stuff, either by our choice or through natural forces that we can't control.

The goal of our planning can be stated simply and clearly: fewer and less. Fewer people, less stuff.

Many people, including many environmentalists we know, prefer not to talk about the growth of the human population as a problem or about population control as a component of a viable environmental policy.[9] Why? Three reasons seem to push people away from this discussion.

The first is that such concerns about population have been associated with a lack of compassion and/or racism, ethnocentrism, and class prejudice.[10] Historically, some of the people who worried about population growth, including Malthus, said some pretty cruel things about poor people and argued that any actions based in benevolence toward the poor would be self-defeating because such measures lead to increased population that would make the situation worse. When a legitimate concern about population turns into a rationalization for harsh social Darwinism, it's not surprising that decent people get nervous about the issue.

Today, some of the most vocal supporters of population control also espouse white supremacist and anti-immigrant sentiments.[11] The ugly history of eugenics lurks in the background as well.[12] As a result, any discussion of population growth as a problem can lead to accusations of bigotry or insufficient understanding of the liberating potential of ecosocialism,[13] which tends to shut down necessary conversations. We are grateful that some environmentalists, such as Eileen Crist, are willing to speak bluntly: "The dismal consequences for Earth and for humanity of an oversized global population are indisputable."[14]

The second reason people might avoid the subject is that no one has ever proposed a viable noncoercive strategy for serious population reduction on the scale necessary, because no such strategy exists. Raising the status of women and educating girls, along with family planning, can reduce birthrates,[15] but not at anywhere near the rate that will be necessary to get to a sustainable population. The most well-known experiment in large-scale limitations on births, China's one-child policy, in place from 1980 to 2016, was controversial on moral and political grounds, and researchers still debate how many births it prevented.[16]

The other side of the population equation—the death rate—is even more vexing. The twentieth century saw declines in infant, child, and maternal mortality, along with the invention of medical technology that could extend people's lives.[17] The question about population reduction requires talking not just about how many kids are born but about how long each person lives, and even fewer people want to talk about that side of the population problem. In the 2009 debate about health insurance and the 2012 presidential election, conservatives whipped up hysteria over "death panels," the argument that moving toward universal health care would result in bureaucrats making decisions about who lives and who dies. The ease with which some politicians were able to scare people with such claims indicates how far the United States is from an honest discussion on the subject of the appropriate level of intervention to prevent death, especially as we age.[18] We need such a debate about setting policy, not only on when to withdraw care from the terminally ill, but also on the wisdom of using a range of life-extending medical procedures (e.g., heart bypasses, organ transplants).

While we need to talk about birth control, just as crucially, we need to talk about whether to continue the current level of death control. In the United States, we ration health care by the ability to pay, which means that those with resources can maximize the use of advanced medical care to extend life. If we were to institute a truly equitable system of health care distribution, the question of death control cannot be avoided. The discomfort with this issue doesn't render the questions irrelevant. As the author of a recent study of human life spans put it, "Many of the key problems we now face as a species are second-order effects of reduced mortality."[19]

Also important to social stability is what is called the dependency ratio, the relationship between people of working age and those who are not working. The youth dependency ratio includes those under the age of fifteen, and the elderly dependency ratio includes those sixty-five and older. A high dependency ratio means working people carry a heavier burden to support those who are not economically active. So if birth-rates were to continue to decline, slowing population growth, and people were to continue to live longer, the dependency ratio will rise over time, with dramatic consequences. In the words of an international team of journalists, "The strain of longer lives and low fertility, leading to fewer workers and more retirees, threatens to upend how societies are organized . . . [and] may also require a reconceptualization of family and nation."[20]

The third reason that people may avoid the population issue is that, whether acknowledged or not, everyone recognizes that raw population numbers are meaningless without attention to per capita consumption, a question that we raised in earlier chapters and will continue to emphasize. That means talking about limits. Imposing limits requires that we distinguish between basic needs (what we truly can't live without, such as food, water, clothing, and shelter), social needs (what's required for humans to flourish, or what we might call enrichment activities such as the arts), and luxuries that we will not be able to support (e.g., not only private jets for the wealthy but also routine air travel for the middle class). It also means talking about redistribution of wealth, not only within societies, but between countries. Those choices require planning within a political process that is committed to the dignity of all people and global solidarity, which requires a willingness to pursue policies with the goal of a rough equality. No such planning has yet happened, and no such political process currently exists. Such a process must start with a commitment to a dramatic reduction in per capita consumption in the developed economies and a recognition that the developing world must abandon the goal of achieving the level of consumption that exists in the affluent countries today. Those goals are not easy to achieve, nor are they fair to everyone, but they are the task ahead.

Those are the impediments, not just to adopting policies, but to even talking about the issue of human numbers and consumption—

what Jackson has long described as the dual population problem: the population of people and the population of people's things.[21] Behind all the denial is the techno-optimism that assumes we will always invent our way out of any problem, which may turn out to be the biggest impediment to meaningful change.

If we could overcome these impediments, what should be our goal? What is a sustainable human population? We don't pretend to know, and failed attempts at prediction in the past have made people understandably wary. But it's safe to say that if our goal is long-term sustainability, the number is well below eight billion people. A lot fewer people, consuming a lot less.

We don't know what the carrying capacity of the planet might have been in the past—before agriculture unleashed the ecological drawdown, before the concentrated assault of the industrial world intensified the degradation of ecosystems. But that's irrelevant because we have to plan for the current degraded state of the world's ecosystems. As a starting point for policy making, we think it makes sense to assume the carrying capacity is no more than half of the current population living with no more than half of the energy and materials the world consumes in the aggregate today. That's a good place to start. Reasonable people with good track records on understanding ecological limits suggest that the human population could stabilize at about two billion.[22] (That was, by the way, the human population in 1927.) One recent analysis concluded that Earth could support three billion people.[23] To get this conversation going in public, what is important is not a specific number but rather a recognition of the scale of change that is needed. Rejecting a growth economy and the irrational consumption of consumer capitalism would make it possible to sustain a human population that is likely no more than half of the current population and likely half of that again.

Finding a humane and democratic path to that dramatically lower number will not be easy. It may not be possible. In fact, if human history is any guide, it's almost certainly not possible. But rather than turn away, we should acknowledge the reality of ecological carrying capacity while pursuing social justice goals and rejecting racist and ethnocentric projects. Refusing to acknowledge difficult problems doesn't allow us to escape them. Instead, denial of reality opens up space for people peddling pseudo-solutions. When reasonable people stay silent, the voices of unreasonable people dominate. Progressives who are unwilling to address the issue of human numbers and consumption cede this terrain to political actors without progressive values who want to use ecological crises to pursue an ugly agenda.[24] To press the question of population and consumption is not reactionary but rather an attempt to forestall reactionary political projects.

We pause here to note the obvious: the analysis just presented is not new, nor are we alone today in this analysis. For a half century, insightful scholars have been making these points. As a young professor, Jackson was thinking about carrying capacity when he proposed a curriculum for "Survival Studies" and collected articles with prescient early warnings in his book *Man and the Environment*,[25] the first edition of which came out in 1971.[26] In the early 1970s, Paul Ehrlich and John Holdren offered the "IPAT" concept to capture the *impact* of human activity on the environment by looking at *population*, *affluence*, and *technology*.[27] In their 1972 book, *The Limits to Growth*,[28] Donella and Dennis Meadows and coauthors used computer modeling to warn that humans were moving beyond Earth's carrying capacity. After years of these authors being dismissed as alarmist, an increasing number of people are recognizing they were right.[29] William Catton, one of the founders of environmental sociology, laid out the argument and evidence regarding overpopulation and consumption in his 1980 book that brought the term "overshoot" to our attention.[30] The ecologists William Rees and Mathis Wackernagel developed ecological footprint analysis to make the unsustainability of contemporary societies easier to grasp, publishing *Our Ecological Footprint* in 1996.[31] Such work continues with scholar-activists such as Richard Heinberg[32] and his colleagues at the Post Carbon Institute.[33]

With these analyses in mind, we reiterate an important point too often overlooked: at the core of what people call "environmental problems"—including, and perhaps especially, climate change—is too many people consuming too much energy. Specific threats to ecosystem health are derivatives of the size of the population and its consumption. There are no long-term solutions to the ecological crisis without coming to terms with that reality. A group of ecologists recently stated the obvious, but such statements need to be repeated: "Large population size and continued growth are implicated in many societal problems."[34]

Scale

At the same time that we move toward a sustainable level of population and consumption, we need to think about the appropriate scale of a human community. What level of social organization is most compatible with a sustainable future? What level of complexity in how we organize our political and economic lives is most likely to get us there? We assume that the coming decades will present new challenges that require people to move quickly to adapt to the fraying and eventual breakdown of existing social and biophysical systems. What ideology and size of governing units is most likely to be workable in a low-energy future? What kind of decision-making processes will be most functional?

Our evolutionary history as a species is relevant in this inquiry. We look to that deeper history not just to remind ourselves of the obvious—that there used to be far fewer people on the planet and fewer human-generated threats to self and other species—but to examine how those people lived. We can't return to the past, but we can ponder what lessons we might take from it.

Let's start with a short recap of human history. The first species in the genus *Homo*, *Homo habilis*, is dated to about 2.5 million years ago, give or take a few hundred thousand years. *Homo sapiens* arrived on the scene sometime between 200,000 and 300,000 years ago. The vast majority of those early humans were mobile, living in small social groups characterized by egalitarian relations that did not result in institutionalized disparities in power, with the exception of those resource-rich ecosystems that allowed some people to amass surpluses and settle in a smaller area. Agriculture first emerged about 10,000 years ago. Domestication arose in multiple places; it was not an overnight "revolution" but a gradual process, with variations in geography and climate determining what plants and animals were available for domestication. From agriculture emerged widespread permanent settlements, large-scale societies, and the domination/subordination dynamic that comes with social hierarchies, starting with patriarchy and men's assertion of control over women's reproductive power and sexuality. In the past 5,000 years, starting in the Bronze and Iron Ages, came what we call civilization: writing, metallurgy, complex societies, bureaucracy, armies, and empires. The past 500 years have seen the rise of nation-states, colonialism, industrial societies, total globalization, biotechnology, digital technology, and an intensification of human-generated ecological crises.

Our point? In evolutionary terms, the large-scale societies created after the introduction of agriculture should be understood as quite strange, in a nonjudgmental sense. The way virtually all people live today is dramatically different from the lifeways of our species through virtually all of our history. That's not because those humans were different from us—genetically, they were pretty much the same—but because they had not pursued agriculture and large permanent settlements or they had not developed the technology to do so.

Our argument is not that agriculture created some new human capacity for dominance or that agriculture developed in the same way everywhere. Human capacities didn't suddenly change with agriculture, and geography would define the parameters within which agriculture developed. Our argument is that some agricultural societies, especially those producing cereal grains, created the conditions for competition to control the surplus, which led to new ideas about ownership and hierarchy. Once the successful carbon seeking of agriculture took hold, the hierarchies it made possible would change the world.

The result is that routine features of the contemporary world, which have existed for barely the blink of an eye in historical terms, would be foreign to every human who lived for about 99 percent of our evolutionary history as a genus and at least 95 percent of our evolutionary history as a species. Superficial cultural practices—what people eat, which musical scale people sing in, the names of the gods they worship—vary from place to place and change over time, of course. But in more profound ways, the past ten thousand years have been unlike all that came before. What we take to be normal in our everyday lives is anything but normal in evolutionary terms.

This is why Jackson has for nearly four decades been suggesting that the key to designing more sustainable systems is to recognize that we are "a species out of context."[35] Evolution by natural selection adapted us to a gathering-and-hunting lifestyle in small band-level societies. Prior to the invention of agriculture, humans were mostly foragers living in a social group of probably no more than one hundred people, and often far fewer. Not only did our bodies evolve for that way of living, but so did our brains (our brains are part of our bodies, of course, but it's important to emphasize this because so many people think of the human mind as somehow being distinct from the body).

We return to a point made in chapter 1: There is such a thing as human nature, just as with all other organisms, and understanding human nature is relevant to designing social systems. Our genetic endowment makes some things possible and some things impossible. No human being can fly in the sense that a bird flies or live underwater in the sense that fish do. But our technological prowess has led people to forget the obvious. Airplanes create the illusion that we can fly and submarines, the illusion we can breathe underwater. Like any organism, we live within not only the biophysical limits of the ecosphere but also the limits of our genetic endowment. That is true not only about our physical capacities but also about our psychological capacities and the social arrangements that work with those capacities.

A caution: Acknowledging this does not mean that we embrace all the claims made in the field of evolutionary psychology, a recent fad in academic psychology, which has put forth a much-debated theory and produced considerable noise as well as some insight.[36] But we recognize the obvious, that evolution is relevant to understanding human psychology. Evolution by natural selection shaped not only the way we walk (bipedalism) but also how we think, feel, and interact (human attitudes, beliefs, and norms of social behavior). The entire organism—our skeletal and muscular systems and our brains—is the product of evolution. The conditions under which we evolved as a species are relevant to understanding ourselves today, a truism that is too often ignored. An earlier flashpoint of this debate was the term "sociobiology," introduced in E. O. Wilson's 1975 book by that name and defined as the "systematic study of the biological basis of all social behavior."[37] Again, we are biological creatures, and it's hard to dispute that there is a biological basis for everything we do (unless one believes in a spiritual plane, with or without a divine force). But just as with evolutionary psychology, it's too easy for claims in sociobiology to run ahead of evidence and offer just-so stories about humans' behavior today.[38]

Research on human social networks suggests that there is a limit on the "natural" size of a human social group, about 150 members. This has been called "Dunbar's number" (after the anthropologist Robin Dunbar),[39] the number of individuals with whom any one of us can maintain stable relationships, which is determined by the size of our brains. We have the cognitive capacity to handle about 150 people in our social groups, and anything larger tends to overload our neural circuits. Moving down from the number 150, Dunbar found that we tend to have groups of about 50 close friends, a tighter circle of about 15 very close friends, and an inner circle of 5 people who provide our most trusted support. We choose our friends, and the norms for friendship vary from culture to culture. However, those patterns are not the product of individual choice or cultural context but rather of human nature. That's the kind of animals we are, based on the size of our brains, which is a product of evolution by natural selection. There are only so many people we can keep track of at one time.

For most of our evolutionary history, humans lived in social groups that matched pretty closely to our cognitive capacity. The kin-based band-level societies of gathering-and-hunting people stayed within Dunbar's number. A band might have connections to other bands, based on shared language and culture. But the development of states and empires is a feature of agricultural societies. Those larger social systems were developed by a species out of context.

Here's an example to illustrate. Jensen regularly taught large university lecture classes of 150 to 300 students, settings in which many students felt uncomfortable as they were forced into close proximity to people they had never met and with whom they may not have had any shared culture. Jensen would ask students to ponder the fact that there were more people in the lecture hall than members of a foraging society might have ever known. "If you feel awkward and don't know exactly what to say to the stranger next to you," Jensen would say, "is that surprising? If you feel uncomfortable, that's normal."

The total undergraduate and graduate enrollment at that university was around 50,000. Most people can't easily find their way in a population of 50,000, or even 300, not because they are odd but because they aren't odd. There are smaller groups on every campus—social clubs, academic organizations, sports teams, performing arts groups—but not every student finds a place in those settings. To help the students who struggle with the scale of the university, administrators created First-Year Interest Groups of about 20 students who took some of their classes together and met weekly with a peer mentor and a university staff member. Although the university didn't use Jackson's phrase, these first-year groups were created as a way to deal with the species-out-of-context problem. The program recognizes that dropping a first-year student into a huge university creates distress for many and that some of those students will believe that it's their fault that they don't fit in rather than understand that the university's structure is the problem.

Another example of this pattern is the megachurch, congregations with thousands of members, which in addition to large spectacle worship services offer myriad specialty groups. In these small groups, members of the congregation build strong connections to a manageable number of people, a cell model of organizing. Megachurch pastors learned something that should not be surprising given our evolutionary history: "The small group was an extraordinary vehicle of commitment."[40] Such models can also be found in political groups, especially in resistance movements worried about being infiltrated by agents of law enforcement or the military. In anarchist organizing on the Left, such small units typically are referred to as affinity groups, "a circle of friends who understand themselves as an autonomous political force."[41]

Not everyone needs the support of small groups to negotiate larger groups of strangers in big institutions. That's not surprising, given that human beings are able to adapt to a wide range of social situations. We are capable of establishing deep, lifelong connections to a small group of family and friends, as well as of finding our place in a city of thousands or even millions of people. But in designing social systems, we should consider what works best over time for the greatest number of people. It makes sense to think about the context in which we evolved as a species. Today we rarely question what we take to be normal ways of organizing human societies, such as nation-states with capitalist economies. But "normal" today is radically different from how we lived for almost all of our evolutionary history: in smaller and much more egalitarian social groups.

If the goal is less hierarchy and more egalitarian relationships, we have to consider the optimal size of social organizations. If we want less hierarchy and meaningful democracy,[42] we should move toward smaller groups. How do we get to that smaller size? Simply declaring that from this point forward everyone in large complex societies shall live in core communities composed of no more than 150 people isn't a viable strategy. First, the majority of people in complex industrial societies have little or no experience with the political dynamics at that level of organization, and the many failed communes are testimony to the fact that the transition is not easy.[43] Second, the existing distribution of wealth and power, concentrated in the corporation and the state, won't disappear simply because some people realize that those institutions are corrosive and unsustainable economic and political formations. The larger society remains, with control of resources and the keys to the armory. Who controls the means of production and who has the most guns matters.

There is no off-the-shelf plan for reaching the most workable level of social organization, just as there is no simple answer to the vexing question of how to reach a sustainable level of the human population. We can start by forming such groups and communities wherever and whenever we have the opportunity,[44] recognizing that such experiments won't turn the tide immediately but create possibilities for the future (more on the idea of a "saving remnant" in chapter 4). No matter how difficult the transition may be, in the not too distant future we will have to live in far smaller and more flexible social organizations than today's nation-states and cities.

Scope

Moving from eight billion people living in large political units to a sustainable human population living in smaller social organizations is a difficult task, unlikely to be achieved by planning in the existing political and economic systems in the time available. There are myriad ways that human-built infrastructure and social systems can—and likely will—fray, falter, fail, and fall apart before we humans can figure out how to manage that task. The unraveling of large systems in the past has generally resulted in social dislocation and intensified conflict, along with lower levels of available material resources and considerable deprivation. There's no reason to assume contemporary societies are immune to similar outcomes as our systems unravel. Anyone from the United States who is reading this has lived through a period of extraordinary material abundance (again, not equally or equitably distributed) and social stability (at the macro level, not necessarily in each person's life or community), which can lead to the assumption that what has been will continue to be. That is not guaranteed, and is increasingly less likely, even in the short term.

That's not a pleasant future to ponder and prepare for, so it's not surprising that many people, especially those in societies whose affluence is based on dense energy and advanced technology, clamor for solutions that claim to be able to keep the energy flowing and the technology advancing. Jackson has long labeled this approach "technological fundamentalism,"[45] a term we introduced in chapter 1. This fundamentalism is a religious-style faith in the ability of societies to solve problems with high energy and high technology, including the problems created by past use of that energy and technology.

Here's an example from *The Restless and Relentless Mind of Wes Jackson*.[46] In the 1980s, the chlorofluorocarbons (CFCs) commonly used in refrigerants and air-conditioning were linked to the growing ozone hole. CFCs were thought to be miracle chemicals when introduced in the 1930s: nontoxic, nonflammable, and nonreactive with other chemical compounds. But while stable in the troposphere, they move to the stratosphere and are broken down by strong ultraviolet light, releasing chlorine atoms that deplete the ozone layer. So international regulations were imposed. CFCs were replaced by the "safer" hydrofluorocarbons (HFCs) and hydrochlorofluorocarbons (HCFCs), until their contribution to global warming was understood,[47] prompting a move to hydrofluoroolefins (HFOs), which are touted as having zero ozone depletion potential (ODP) and low global warming potential (GWP).[48] Miracle Chemical #1 is introduced but later found to cause problems and banned, replaced by Miracle Chemical #2, which is later found to cause problems and banned, replaced by Miracle Chemical #3 . . . Lather, rinse, repeat.

Such hubris flows from an unwarranted confidence in the ability of humans to understand complex questions definitively. Because humans in the past century have been able to transcend temporarily the biophysical limits of the ecosphere, this faith in technology allows people to avoid a simple reality: an economics of endless growth on a finite planet will end badly. Because the obsession with growth defines virtually all economies on the planet, the bad ending will not be contained to specific societies but will be global. The response of the technological fundamentalists is simple: we need not curb our aggregate consumption as a species because we will invent whatever we need in a "green-energy cornucopia,"[49] which some Cornucopians argue should include nuclear power (more soon on why that's a bad idea).[50]

The irrationality of this temporary evasion of the biophysical limits of the planet is captured in an old joke. A man (given the patriarchal foundation of unsustainable systems, the sex-specific "man" is appropriate here) jumps off a hundred-story building and, as he passes each floor, says to himself, "So far, so good." The experience of falling can be exhilarating, but at some point, the man will hit the pavement. Technological fundamentalists seem to believe that we can remain in free-fall forever through innovation. And if human societies were to ever get close to the pavement, then innovation will provide a giant net to catch us. These faith-based beliefs require us to assume not only that new inventions will solve our not yet solved problems, but that we can continue to manage the technology we have already created without catastrophe.

Those betting on technology to forestall the consequences of biophysical limits believe that humans are competent to manage not only our current interventions into the larger living world but also the more dramatic interventions that likely will be necessary in the future as old technologies fail to meet their promise. The ultimate example of this runaway hubris is geoengineering, the manipulation on a global scale of ecological processes to counteract the effects of global warming and respond to other ecological crises.[51] Rather than accept limits and reduce energy consumption, technological fundamentalists propose schemes such as releasing sulfate aerosol particles into the stratosphere to reflect more sunlight back to space. If that sounds crazy, maybe it is.[52] But lots of not-crazy people are willing to double down on even more extensive human manipulation of the ecosphere.[53]

The authors are convinced that is the wrong approach. The evidence suggests that humans are far less competent than the dominant culture assumes, cannot successfully manage the current infrastructure of the industrial world much longer, and will not have any greater success managing the imagined technologies of the future.

An important illustration of this is the transformation of agriculture in the post–World War II era, which is seen by many as a great triumph. The dominant culture's story of the Green Revolution goes something like this: in the face of worldwide hunger and the threat of social destabilization, agronomic research created new and improved versions of staple crops, farming methods, and technologies that saved a billion people from starvation. The leading researcher in this area—Norman Borlaug, often called "the father of the Green Revolution"—won the Nobel Peace Prize in 1970 for providing "bread for a hungry world." This application of human ingenuity dramatically increased yields through hybridized seeds, increased irrigation, modern farm management, and synthetic fertilizers, pesticides, and herbicides. What more could we ask for?

A more complete assessment would include the negative ecological and social consequences of this high-energy/high-technology approach to agriculture.[54] Chemical contamination of soil and water increased. Soil erosion continued. Use of fossil fuels increased, as did the environmental degradation from extraction of that energy and climate change when it was burned. One-size-fits-all agronomic approaches that worked when implemented with petrochemical inputs and cheap energy (cheap, that is, if costs are calculated only in the short term) replaced locally adapted methods that typically used much less fossil energy and fewer or no chemicals. Peasant farmers were persuaded, or forced, to adopt these new "scientific" methods, which undermined their independence and resilience. And without a commitment to economic justice, more food on the market didn't always feed hungry people who had little money. First world dominance was strengthened at the cost of greater third world dependency. Third world agriculture moved toward being as brittle as first world farming.

Agribusiness executives and policy makers rarely talk about just how insecure agricultural production has become, how dependent humanity is on a brittle system. Consider this: the increase in the human population from two billion in 1927 to nearly eight billion today was made possible by yield increases that are primarily the result of the industrial production of anhydrous ammonia as a source of nitrogen fertilizer for modern agriculture. The Haber-Bosch process, invented in the early twentieth century, typically uses natural gas as the feedstock to turn tight-bonded atmospheric nitrogen into ammonia.[55] This industrial process "solved" the problem of soil nitrogen fertility and declining supplies of natural fertilizers such as guano. Scholars estimate that this industrial process supports nearly half the world's population.[56] A disruption in the supply of energy needed for this industrial process would mean a worldwide famine of unprecedented proportions. Meanwhile, the widespread use of these fertilizers means that after being spread over millions of acres of grain-producing fields, the surplus industrial nitrogen finds its way into waterways until it meets the ocean waters, where it creates huge dead zones. On the way downriver, cities spend millions of dollars to remove it from drinking water, in some places failing so dramatically that people have to drink bottled water.

The technocrats and politicians who gave us the Green Revolution believed that managing this industrial style of agriculture within an increasingly globalized economic system was within the scope of their competence. The experience of the past seventy-five years suggests just the opposite. We do not doubt those people's intellectual abilities. For purposes of this discussion, we accept their stated goal of wanting to feed hungry people, though some of those politicians also were worried about the kinds of radical revolutions that hungry people would support. Yields increased, but the health of ecosystems and the resilience of local economies declined. The agricultural experts who gave us the Green Revolution could not successfully manage the complex systems they created, yet many of those same experts continue to press forward with more high-energy "solutions" to the problem of feeding an expanding population on rapidly degrading soils.

The Green Revolution is one of many examples of the inability of human intelligence to predict all the consequences of technologies and policies. But technological fundamentalists never stop believing that the use of high-energy advanced technology is always a good thing. Nor do they seem to doubt that the continued use of ever more sophisticated high-energy advanced technology can solve our problems, including the problems caused by the unintended consequences of earlier technologies. These fundamentalists act as if human knowledge is adequate to run the world. But to claim such abilities, we have to assume we can identify all the patterns in nature and learn to control all aspects of nature. That we so clearly cannot do those things does not disturb technological fundamentalists' faith.

Nuclear power is another example of this misplaced faith in humans' ability to control our interventions. In 1979, one of the reactors at the Three Mile Island Nuclear Generating Station in Pennsylvania partially melted down as a result of equipment malfunctions and operator errors, exacerbated by design flaws. Not surprisingly, that scared a lot of people, but officials told us that we need not worry about future accidents because of improvements in emergency response plans, operator training, engineering, and radiation protection. The reactor meltdown in northern Ukraine at the Chernobyl Nuclear Power Plant in 1986 scared a lot of people too, but nuclear energy supporters assured us that the problem was bad reactor design and that design improvements had eliminated the possibility of a repeat of that catastrophe. The radiation leaks from Japan's Fukushima Daiichi Nuclear Power Plant in 2011 after an earthquake and tsunami *really* scared a lot of people. But nuclear supporters once again studied the problems and once again counseled improvements in systems, resources, and training, along with the recommendation to "strengthen capabilities for assessing risks from beyond-design-basis events."[57] Experts use the term "beyond design basis" to refer to accidents that were judged to be too unlikely and therefore were not fully considered in the design process. Very reassuring.

These lessons learned from unanticipated accidents at three different nuclear power plants should reassure us of what? That future unanticipated problems with other nuclear power plants won't be serious? That engineers have acquired clearer foresight and the potentially catastrophic risks of nuclear power generation have been eliminated? That once utility companies see that the goal of maximizing profit has led to unacceptable levels of risk they will stop maximizing profits? That complex systems that create risk can be made safe with more complexity? That there will never be another "normal accident,"[58] the term coined for failure that can be understood in hindsight but cannot be predicted because of the complexity of a system? And don't forget that the problem of the safety of nuclear waste storage has never been solved. We've made our point.

Back to the joke about the hundred-story building. It would have been wiser never to have jumped off the building, better never to have rushed headlong into an uncritical expansion of the human population through an embrace of dense energy and the industrial worldview. But once we jumped, we have no choice but to continue to use our knowledge to try to create a soft landing. But that suggests—and this stretches the metaphor—that on the way down we should not add more technology unless it will contribute directly to making a soft landing as soft as possible. We are not arguing against research and development of renewable energy, for example. We are arguing that such energy should be understood as necessary to help us power down, not prop up existing levels of consumption.[59]

But to increase our chance of success in this use of knowledge, we need humility. Another idea Jackson has promoted for decades is an ignorance-based worldview, which is not a call to being stupid but instead a reminder that the limits of human knowledge should curb our hubris.[60] Human knowledge has expanded dramatically since the Enlightenment and the Scientific Revolution, especially during the high-energy industrial era. Not all of that knowledge has been destructive, and much of it has enriched our lives. But what we don't know still far outstrips what we do know, and always will. We might want to keep that in mind as we go forward producing new knowledge. Human capabilities should always be deployed with an eye on human fallibility. We need to use technology judiciously, in a way that allows us to adjust to our inevitable mistakes rather than magnify the consequences of those mistakes. We need to be better students of the exits so that when things go wrong, as they always do and always will, we can find our way out.

Speed

If there is to be a decent human future, we face the tasks of reducing the number of humans and aggregate consumption, moving to smaller and more flexible political units and social organizations, and recognizing the limits of our intelligence to manage complex systems. That leaves the question of speed: how fast must we move toward these dramatically different living arrangements and collectively self-imposed limits if we are to avoid catastrophic consequences?

We have already acknowledged that it is folly to offer precise predictions, given human cognitive limits and the complexity of the world. But this does not mean we should abandon attempts to understand the trajectory of human societies or stop trying to deepen our understanding of where we are heading. A good example is the debate over peak oil, the point at which humans will have extracted about half of all oil that will ever be extracted. After that point, petroleum extraction will permanently decline. For decades, the common wisdom was that the United States had hit that peak in 1970, but new sources of hydrocarbons opened up with the expanded use of hydraulic fracturing, commonly known as fracking. Modern fracking began in the 1940s but really took off in the 1990s when new methods for horizontal drilling were developed.

Oil is a nonrenewable resource, a finite resource. But when we hit the peak depends not only on how much oil exists underground but also on extraction technology, government policy, consumer demand, market reactions, and competing fuel sources. Still, even if we don't know when, we know we will run out of oil that can be easily extracted. We also know that the ecological costs of extracting hydrocarbons that are more difficult to get and process, such as the tar sands of Canada, are unacceptable, especially given the low energy return on investment (EROI). That's the ratio of the amount of energy obtained from a resource to the amount of energy needed to produce it (sometimes called EROEI, energy returned on energy invested).

In short, it seems clear that even if greenhouse gases weren't a problem, the era of cheap oil is over and we can assume the end of the era of oil is coming. We don't need exact predictions to assess the trajectory and act on that assessment. And, remember, if we factor in the ecological damage from extracting and burning all that oil, then "cheap oil" should always be in scare quotes, since it was never really cheap.

Looking at one variable, such as oil extraction, is vexing enough. Making any prediction about how complex global systems will play out is just plain silly. But we can offer honest, informed speculation based on history and biology—our best guess on which we can base planning. Without claiming to be clairvoyant, we feel confident in saying this, bluntly: Individuals and societies must enact significant changes much faster than we have been willing to do so far and faster than it appears that the human species has been capable of acting to date, perhaps faster than we ever will be able to act.

Not everyone agrees with that assessment, of course; different readings of history and biology are possible.[61] We welcome conversation with critics but suggest they defend those different readings with substantive arguments, avoiding clichés and slogans. There are lots of sayings in various cultures that emphasize the need for patience and perseverance, which are positive qualities so long as they are grounded in reason and evidence. One such saying, usually credited as a Chinese proverb, asserts, "The best time to plant a tree was twenty years ago. The second-best time is now." It's true enough that just because one should have started working toward a goal earlier is not a reason to never begin the work. But it's also true that second-best action might not be enough to accomplish the original goal, which means expectations have to change.

Since Jensen recently began tending apple trees in New Mexico, we offer this analogy. Let's say it takes the variety of apple tree you are planting five years to grow to the point that it will bear fruit, and you are planning an apple festival for next year. In that case, this cliché would ring hollow: "The best time to plant an apple tree was five years ago. The second-best time is now." Expectations have to change. It's best to give up on the festival next year. Continuing to plan for that event with the hope that the tree will produce the apples would not be evidence of patience and perseverance but of being slow-witted. Either find another source for the apples or cancel the festival. Perhaps a techno-optimist would suggest that our desire to have apples will incentivize people in the market to create a new way to get the apples sooner than natural processes permit. Plant breeders can do some pretty amazing things, but it takes time. More time, in this case, than is available.

At the risk of offending tree planters, we want to push the analogy in another direction. When people say that "the second-best time is now," they are clearly wrong. If we proceed in one-year increments from twenty years ago, the second-best time would have been nineteen years ago. The reason for being a stickler on this point is to force us to investigate the reasons that, in this metaphor, tree planting kept being delayed. In the case of the ecological crises we face, the warning light has been flashing red not for years but for decades, in some ways even for centuries. If year after year the evidence piled up and still there was no meaningful collective action to deal with questions of size, scale, and scope, what makes us believe that piling more evidence in business-as-usual fashion will produce the change we need at the speed required?

Our analogy is not an argument for inaction but rather a suggestion that actions we take should be consistent with our best assessment of reality. And in this case, we assert that policies based on the belief that eight billion people will continue to live in high-energy industrial societies are unwise, a bit like planning for an apple festival when there won't be any apples. It's unwise to believe that what is often called the magic of modern science and engineering can actually work magic. It's unwise to think that a species that has resisted taking collective action at the speed necessary will find even more rapid and dramatic change easy in the future.

Our critics might point to examples of societies instituting needed but unexpected change quickly, such as the United States shifting much of its industry from civilian to military production as it entered World War II. But that example highlights how a society can react to tangible, immediate threats by redirecting resources within the existing industrial system to new priorities and says nothing about the much greater challenge of remaking societies to be more ecological on a global scale. Although retooling American industry virtually overnight was no small task, it was relatively simple compared to the changes in economics, politics, and culture that are necessary if we are to put the human species on an ecologically sustainable trajectory.

Threat Perception and Risk Assessment

It's often observed that humans evolved to perceive and act on visible, immediate threats in our environment, not on complex, hard-to-identify, long-term threats on a global scale. We might call this the present-and-visible bias in risk assessment. Combined with our inability to predict the unintended consequences of new technologies and the unplanned outcomes of human actions in complex systems, this bias leads to unrealistic assessments of the time frame available for change. Collectively, we have a false sense of the control that we have over the larger living world, not only what can be accomplished, but how quickly we can accomplish it. This is an observation not about humans today but about humans. A leading paleoanthropologist puts it this way:

> Apart from death, the only ironclad rule of human experience has been the Law of Unintended Consequences. Our brains are extraordinary mechanisms, and they have allowed us to accomplish truly amazing things; but we are still only good at anticipating—or at least of paying attention to—highly immediate consequences. We are notably bad at assessing risk, especially long-term risk. We

> believe crazy things, such as that human sacrifice will propitiate the gods, or that people are kidnapped by space aliens, or that endless economic expansion is possible in a finite world, or that if we just ignore climate change we won't have to face its consequences. Or at the very least, we act as if we do.[62]

Such warnings may seem like one more chorus of "the sky is falling" from overly dramatic environmentalists. But when research from the insurance industry starts sounding like it came from environmental groups, perhaps we should pay attention. "A fifth of countries worldwide at risk from ecosystem collapse as biodiversity declines," reports Swiss Re Group, a leading reinsurance firm (the firms that regular insurance companies call when they need to lay off some of the risk in policies they have written). The company's study "highlights the dangers of these economies potentially reaching critical tipping points when essential natural resources are disrupted."[63] A year later, another Swiss Re report predicted that on the currently anticipated trajectory, the world could lose up to 10 percent of total economic value by midcentury from climate change.[64]

Other business analysts agree. With the understated tone that one expects from a major worldwide accounting-consulting firm, we learn, "Climate risks present a significant challenge for the insurance industry, since they are likely to increase over time. Finding a balance between ensuring affordability and availability and managing financial stability may get tougher for insurers if extreme weather conditions continue to escalate."[65]

Here's a similar assessment from the Commodity Futures Trading Commission, a government agency not known for extremism. A 2020 report warns that the entire financial system could be upended as a result of climate change alone.

> Climate change poses a major risk to the stability of the U.S. financial system and to its ability to sustain the American economy. Climate change is already impacting or is anticipated to impact nearly every facet of the economy, including infrastructure, agriculture, residential and commercial property, as well as human health and

> labor productivity. Over time, if significant action is not taken to check rising global average temperatures, climate change impacts could impair the productive capacity of the economy and undermine its ability to generate employment, income, and opportunity. Even under optimistic emissions reduction scenarios, the United States, along with countries around the world, will have to continue to cope with some measure of climate change–related impacts.[66]

We aren't economists, auditors, consultants, or CEOs. We don't have access to proprietary research that corporations often commission, and we assume that the risk-assessment experts are not putting everything they know on the company website. But one need not read beyond the headlines to ask, when will weaknesses in a system threaten business as usual? When will the ecological crises start tanking not only the insurance industry but the whole economy too, judged by traditional measures of profit and loss? We don't know the answers. But we assume that by the time corporate quarterly reports start signaling such problems, it will be too late.

One last quote from folks not generally assumed to be in the pocket of the environmental movement, the US "intelligence community." The director of National Intelligence's annual threat assessment for 2019 summed it up this way:

> Global environmental and ecological degradation, as well as climate change, are likely to fuel competition for resources, economic distress, and social discontent through 2019 and beyond. Climate hazards such as extreme weather, higher temperatures, droughts, floods, wildfires, storms, sea level rise, soil degradation, and acidifying oceans are intensifying, threatening infrastructure, health, and water and food security. Irreversible damage to ecosystems and habitats will undermine the economic benefits they provide, worsened by air, soil, water, and marine pollution.[67]

We offer a conclusion that is well supported by the extensive evidence that is publicly available. The modern economy created by the Industrial and Digital Revolutions is not *running* out of time but rather *has run* out of time to correct the course. We should not be thinking about how to change our practices to allow this world to continue with nearly eight billion people with varying levels of access to high-energy/high-technology living. Instead, we should be planning for fewer people and less consumption, in smaller social and political organizations, with greater humility about tinkering with our surroundings.

We should have thought about this 10,000 years ago when we tinkered with ecosystems in early agriculture and started the long process of exhausting soil carbon, but we didn't. We should have thought about this 5,000 years ago when we learned to smelt metals and started the long process of exhausting the carbon of forests, but we didn't. We should have thought about this 250 years ago when the steam engine ushered in the Industrial Revolution and we started exhausting the carbon of fossil fuels, but we didn't. We should have thought about this 100 years ago when new forms of mass media made it frighteningly easy for concentrated power to shape the public mind and transform our carbon-seeking instincts into the madness of consumer capitalism, but we didn't.

Hard questions lead to painful conclusions. We are starting too late to prevent billions of people from enduring incalculable suffering. We are starting too late to prevent the permanent loss of millions of species and huge tracts of habitat. We are starting too late, but we have to start. How should we understand this work? Honestly. We believe that we start by telling ourselves and each other the truth.

THREE

WE ARE ALL APOCALYPTIC NOW

In the late 2000s, Jensen was getting ready to leave an auditorium after delivering a lecture in which he had described the multiple cascading crises that human societies should be addressing but instead were either ignoring or downplaying. A man from the audience lingered, following Jensen down the aisle to ask, "Do you think society is going to collapse?"

"Well, this society is based on an unsustainable extraction of resources to support unsustainable living arrangements," Jensen said, "so we can't expect it to continue this way indefinitely."

"When will that happen? When will it fall apart?"

"There's no way to predict," Jensen replied, "because there are too many contingencies, and the failures don't happen all at once."

But the man pressed: "What's your best guess?"

"I don't like to guess."

"I know, but you must have thought about it," he said.

"Yes, it's hard not to think about it," Jensen finally answered. "I work on the assumption that if not within my lifetime, it's likely coming within the lifetime of my child [in middle school at the time]."

"That's pretty apocalyptic," the man responded.

"We are all apocalyptic now," Jensen said.

Jensen has been using that phrase ever since to suggest that people who pay attention to the data on ecological realities—whether or not they acknowledge it to others—are thinking apocalyptically or will be thinking that way soon enough.[1] We bet that even Lawrence Summers, who has preached the gospel of growth for decades, might be rethinking the statement he made in 1991, when he was chief economist at the World Bank:

> There are no limits to the carrying capacity of the Earth that are likely to bind any time in the foreseeable future. There isn't a risk of an apocalypse due to global warming or anything else. The idea that the world is headed over the abyss is profoundly wrong. The idea that we should put limits on growth because of some natural limit, is a profound error and one that, were it ever to prove influential, would have staggering social costs.[2]

The staggering social costs are with us, not because we put limits on economic growth but because we failed to do so, because we were slow to embrace an apocalyptic worldview.

We Are All . . .

The first use of the formulation, "We are all —— now," appears to have been in 1887, when William Harcourt of the British Liberal Party responded to the accusation that proposed legislation was socialist by suggesting that we are all socialists now. Even capitalists, he was pointing out, have to recognize the obligation of the government to provide some level of support for people.

In 1965, the conservative libertarian economist Milton Friedman was quoted as saying, "We are all Keynesians now,"[3] to acknowledge the dominance of the economist John Maynard Keynes's ideas about the government's role in managing demand, and a few years later Richard Nixon picked up the idea. The phrase gradually entered the everyday lexicon through increasingly wide uses. For example, after the terrorist attacks of September 11, 2001, many non–New Yorkers said, "We are all

New Yorkers now." Our favorite use of the phrase comes from *Genius: The Life and Science of Richard Feynman*, even though the lineage from Harcourt isn't clear. The book's author, James Gleick, reports that the first jubilant reaction of the scientists to the successful test of the atomic bomb eventually shifted, and more of them reported having had doubts.

> [Los Alamos Laboratory Director Robert] Oppenheimer, urbane and self-torturing aficionado of Eastern mysticism, said that as the fireball stretched across three miles of sky . . . he had thought of a passage from the Bhagavad Gita, "Now I am become Death, the destroyer of worlds." The test director, Kenneth Bainbridge, supposedly told him, "We are all sons of bitches now."[4]

Many discussions of the apocalyptic in contemporary US culture focus on the Book of Revelation, also known as the Apocalypse of John, the final book of the Christian New Testament. The two words are synonymous in their original meaning: *revelation* from Latin and *apocalypse* from Greek both mean a lifting of the veil, a disclosure of something hidden from most people, a coming to clarity. But in popular culture, "apocalypse" has come to mean the end of the world.

To restate our position: We do not believe the world will end. Human abuse of ecosystems cannot destroy Earth. Nor do we expect the human species to become extinct anytime soon, although extinction is the fate of most species. But both of these meanings—today's common usage about endings and the original Greek about insights and clarity—are helpful in focusing our attention.

First, while the end of the world is likely not at hand (at least not until the sun burns out in several billion years), some things will end, such as the unsustainable and unjust economic, political, and cultural systems that currently dominate human societies. If we accept that those systems are, in fact, unsustainable, then it's a matter of simple logic: an unsustainable system cannot be sustained, at least not indefinitely and not without serious modification.

Second, to cope with that understanding of apocalypse, we need to lift the veil, to be able to see reality as clearly as possible. Our chances of coping successfully with the "end times" of those human-created systems increase if we are diligent in learning how the laws of physics and chemistry, along with the lessons of biology, are relevant to our struggles. The lesson that many religious believers draw is that no human power is greater than God, an idea we would secularize: no human system can ignore the forces of the larger living world, which are far more powerful than we are.

To make sure that we are not misunderstood, let us restate this. Invoking the apocalyptic is, for us, not a theological exercise. We aren't interested in arguments among some Christians about premillennialism versus postmillennialism (whether Jesus will come before or after the predicted Millennium period of peace on Earth), or any other esoteric debates over eschatology (the ultimate fate of human souls and the created order). But we find value in the insights of theologians who work with those religious texts and traditions to help us understand contemporary conundrums.

In this approach, we aren't suggesting that Christianity is the only vehicle for such conversations; obviously, many other cultural frameworks are relevant. We focus on the Christian tradition, first, because it is the one in which we were raised and we feel a right to make a claim on its stories, to interpret them in a progressive and ecological fashion that challenges the dominant culture's interpretations; and second, because, for better or worse, it was the Christian nations of Europe that harnessed technologies to use power in a way that allowed them to dictate much of the map of the world we live in, with those inequalities of wealth. That leaves Christianity as not only the largest world religion in number of adherents but also the most influential. So the struggle to interpret Christian stories is important, independent of anyone's background. Toward that end, we will embrace a version of the apocalyptic from the Christian tradition, with our own spin, of course.

Systems of Power: Royal, Prophetic, Apocalyptic

As we've pointed out, both of us have spent much of our adult lives engaged in intellectual work (research in the library and the field, along with writing and teaching). We have been employed in institutions (media outlets, universities, nonprofit organizations) that are funded by various sectors of the dominant culture (corporations, governments, foundations, rich people). We have had a lot of time to read, think, and talk, all of which has been subsidized by others. Both of us are grateful for what we've been allowed to do.

What kind of intellectuals do we strive to be? What is our orientation to the dominant culture that we critique but that also has funded us? Here, we find useful guides in the Hebrew and Christian Bibles, what we grew up calling the Old and New Testaments. Those writings are complex, of course, but we highlight three different relationships to systems of power: royal, prophetic, and apocalyptic. We see lessons to be learned in those texts, aided by some critically minded theologians who have been unafraid to confront themselves and the dominant culture.

We use the term "royal" not to describe a specific form of executive power but as a critique of a system that concentrates authority and marginalizes the needs of ordinary people. The royal tradition, in this context, describes ancient Israel, the Roman Empire, European monarchs, or the contemporary United States—societies in which those holding wealth and power can ignore the needs of the majority of the population if they so choose, societies in which the wealthy and powerful tend to offer pious platitudes about their beneficence as they pursue policies to enrich themselves.

The theologian Walter Brueggemann identifies this royal consciousness in ancient Israel as the period in which Solomon strayed from the wisdom of Moses. Affluence, oppressive social policy, and a static religion transformed a God of liberation into one used to serve an empire. This corrosive consciousness develops not only in top leaders but throughout the privileged sectors, often filtering down to a wider public that accepts a power system and its cruelty: "The royal consciousness leads people to numbness, especially to numbness about death."[5]

The inclusion of the United States in a list of royalist societies may seem odd, given the country's democratic traditions (however frayed), but it is a nation that has been at war—either shooting wars or cold wars for domination—for our entire lives. Economic inequality and the resulting suffering have deepened in our lifetimes, facilitated by a government so captured by concentrated wealth that attempts to renew the moderate New Deal–era social contract seem radical to many. Brueggemann describes such a culture as one that is "competent to implement almost anything and to imagine almost nothing,"[6] though an empire's competence wanes over time, a process visible in the United States today. Much of the intellectual establishment—not just the right wing but also centrists and liberals—either explicitly endorses or capitulates to royal power. Scripture makes it clear that such royal traditions are problematic and must be challenged by those speaking prophetically, a tradition that requires some definition.

Remember what real prophets are not. They do not predict the future or demand that others follow them in lockstep. In the Hebrew and Christian traditions, prophets are the figures who remind the people of the best of the tradition and point out how the people have strayed. Cultivating a prophetic imagination and speaking in a prophetic voice to call out injustice requires no special status in society, and no sense of being special. Claiming the prophetic tradition requires only honesty and courage, a willingness not only to confront the abuses of the powerful, but to acknowledge our own complicity. To speak prophetically requires us first to see the world accurately—both how political and economic systems create unjust and unsustainable conditions and how we who live in the privileged parts of the world are implicated in those systems.

The Hebrew Bible offers us many models. Amos and Hosea, Jeremiah and Isaiah—all rejected the pursuit of wealth or power and argued for the centrality of kindness and justice. The prophets condemned corrupt leaders but also called out privileged people in society who had turned from the demands of justice, which the faith makes central to human life. In his analysis of these prophets, the scholar and activist Rabbi Abraham Joshua Heschel concluded:

> Above all, the prophets remind us of the moral state of a people: *Few are guilty, but all are responsible.* If we admit that the individual is in some measure conditioned or affected by the spirit of society, an individual's crime discloses society's corruption.[7]

Brueggemann encourages preachers to think of themselves as "handler[s] of the prophetic tradition," which doesn't mean imposing their views and values on others but being willing to "connect the dots."

> Prophetic preaching does not put people in crisis. Rather it names and makes palpable the crisis already pulsing among us. When the dots are connected, it will require naming the defining sins among us of environmental abuse, neighborly disregard, long-term racism, self-indulgent consumerism, all the staples from those ancient truth-tellers translated into our time and place.[8]

Invoking the prophetic in the face of royal consciousness does not promise quick change and a carefree future, but it implies that a disastrous course can be corrected. But what if the justification for such hope evaporates? When prophetic warnings have been ignored, time and time again, what comes next? That is when an apocalyptic sensibility is needed.

Again, to be clear: "apocalypse" in this context does not mean lakes of fire, rivers of blood, or bodies raptured up to heaven. The shift from the prophetic to the apocalyptic can instead mark the point when hope for meaningful change within existing systems is no longer possible and we must think in dramatically new ways. Invoking the apocalyptic recognizes the end of something. It's not about rapture but a rupture severe enough to change the nature of the whole game.

The prophetic imagination helps us analyze and strategize about the historical moment we're in, typically with the hope that the systems in which we live can be reshaped to stop the worst consequences of the royal consciousness, to shake off that numbness of death in time. When that no longer seems possible, then it is time to think about what's on the other side. Because no one can predict the future, these two approaches are not mutually exclusive. People should not be afraid to think prophetically and apocalyptically at the same time, as did some of the prophets of old. For example, chapters 24 through 27 of the Book of Isaiah are routinely referred to as "Isaiah's Apocalypse."[9] We can simultaneously explore immediate changes in the existing systems to ameliorate suffering as best we can while thinking about new systems for the future.

In a culture that encourages, even demands, optimism no matter what the facts, it is important to consider plausible alternative endings. Anything that blocks us from looking honestly at reality, no matter how harsh the reality, must be rejected. To borrow an often-quoted line of James Baldwin, "Not everything that is faced can be changed; but nothing can be changed until it is faced." The line is from an essay titled "As Much Truth as One Can Bear" about the struggles of artists to help a society, such as white supremacist America, face the depth of its pathology. Baldwin, writing with a focus on relationships between humans, suggested that a great writer attempts "to tell as much of the truth as one can bear, and then a little more."[10]

We would take Baldwin a step further. To speak from the royal tradition is to tell only those truths that the system can bear. To speak prophetically is to tell as much of the truth as one can bear and then a little more. To speak apocalyptically is to tell as much of the truth as one can bear, then a little more, and then all the rest of the truth, whether one can bear it or not.

If it seems like all the rest of the truth is more than one can bear, that's because it is. We are facing new, more expansive challenges than ever before in history. Never have potential catastrophes been so global. Never have social and ecological crises of this scale threatened at the same time. Never have we had so much information about the threats that we must come to terms with. If that seems overwhelming, that's because it is overwhelming. No one living at this moment in history—including the two of us—can really bear all of the truth. But we stand a better chance of fashioning a sensible path forward if we could all adopt an apocalyptic sensibility and, collectively, try to help each other bear all of that truth.

Objections and Obfuscations

It's time for a summary of our assessment. The human species faces multiple cascading social and ecological crises that will not be solved by virtuous individuals making moral judgments of others' failures or by frugal people exhorting the profligate to lessen their consumption. Things are bad, getting worse, and getting worse faster than we expected. This is happening not just because of a few bad people or bad systems, though there are plenty of people doing bad things in bad systems that reward people for doing those bad things. At the core of the problem is our human-carbon nature, the scramble for energy-rich carbon that defines life. Technological innovations can help us cope but cannot indefinitely forestall the dramatic changes that will test our ability to hold onto our humanity in the face of dislocation and deprivation. Although the worst effects of the crises are being experienced today in developing societies, more affluent societies aren't exempt indefinitely. Ironically, in those more developed societies with greater dependency on high energy and high technology, the eventual crash might be the most unpredictable and disruptive. Affluent people tend to know the least about how to get by on less.

When presenting an analysis like this, we get two common responses from friends and allies who share our progressive politics and ecological concerns. The first is the claim that fear appeals don't work. The second is to agree with the assessment but advise against saying such things in public because people can't handle it.

On fear appeals: The reference is to public education campaigns that seek, for example, to reduce drunk driving by scaring people about the potentially fatal consequences. Well, sometimes fear appeals work and sometimes they don't,[11] but that isn't relevant to our point. We are not focused on a single behavior, such as using tobacco, nor are we trying to develop a campaign to scare people into a specific behavior, such as quitting smoking. We are not trying to scare people at all. We are not proposing a strategy using the tricks of advertising and marketing (the polite terms in our society for propaganda). We are simply reporting the conclusions we have reached through our reading of the research and personal experience. We do not expect that a majority of people will agree with us today, but we see no alternative to speaking honestly. It is because others have spoken honestly to us over the years that we have been able to continue on this path. Friends and allies have treated us as rational adults capable of evaluating evidence and reaching conclusions, however tentative, and we believe we all owe each other that kind of respect.

We are not creating fear but simply acknowledging a fear that a growing number of people already feel, a fear that is based on an honest assessment of material realities and people's behavior within existing social systems. Why would it be good strategy to help people bury legitimate fears that are based on rational evaluation of evidence? As Barbara Ehrenreich points out, an obsession with so-called positive thinking not only undermines critical thinking but also produces anxiety of its own.[12] Fear is counterproductive if it leads to paralysis but productive if it leads to inquiry and appropriate action to deal with a threat. Productive action is much more likely if we can imagine the possibility of a collective effort, and collective effort is impossible if we are left alone in our fear. The problem isn't fear but the failure to face our fear together.

On handling it: It's easy for people—ourselves included—to project our own fears onto others, to cover up our own inability to face difficult realities by suggesting the deficiency is in others. Both of us have given lectures or presented this perspective to friends and been told something like, "I agree with your assessment, but you shouldn't say it publicly because people can't handle it." It's never entirely clear who is in the category of "people." Who are these people who are either cognitively or emotionally incapable of engaging these issues? These allegedly deficient folks are sometimes called "the masses," implying a category of people not as smart as the people who are labeling them as such. We assume that whenever someone asserts that people can't handle it, the person speaking really is confessing, "I can't handle it." Rather than confront their own limitations, many find it easier to displace their fears onto others.

We may not be able to handle the social and ecological problems that humans have created, if by "handle" we mean considering only those so-called solutions that allow us to imagine that we can continue the high-energy/high-technology living to which affluent people have become accustomed and to which others aspire. But we have no choice but to handle reality, since we can't wish it away. We increase our chances of handling it sensibly if we face reality together.

Is Collapse Coming?

We realize that "apocalypse" for many people has become synonymous with a reactionary theology that turns a book of the Christian Bible into the handbook for a death cult's predictions of horrors to come and fantasies about a magical deliverance for the righteous. That's not us.

We are focused on the evidence about the biophysical realities we need to face, paying close attention to the assessments of people with relevant expertise. For example, a survey of more than two hundred scientists from fifty-two countries who study global change highlighted a set of global risks—climate change, extreme weather, biodiversity loss, food crises, and water crises—that could lead to a global systemic crisis.[13] Their conclusion: "Together these five risks threaten the continued integrity of the biosphere and its capacity to support itself and human life." Increasingly, scientists are speaking of the connection between ecological crises and an unsustainable economic system.[14]

With assessments like this stacking up, it's irresponsible not to ask hard questions. Are the political and economic systems that shape the contemporary world consistent with a sustainable large-scale human presence on the planet? Does mass-consumption consumer capitalism premised on endless growth have a future? How long can a high-energy/high-technology system that draws down the nonrenewable ecological capital of the planet expect to continue?

We are apocalyptic; we think modern systems are coming to an end, and we need to lift the veil that obscures an honest assessment of what those end times will require of us. But we are not scholarly "collapsologists,"[15] a term that has been used by some to describe an emerging research community studying systemic risk in industrial society. We have great interest in the question but have not studied the history of societal collapses in the kind of depth that might allow us to make predictions,[16] which we don't like making anyway. We also are not practicing "collapsonauts"[17] in our daily lives, in the sense of belonging to a group that is creating alternative ways of living "off the grid."[18] We try to be frugal but haven't significantly changed our own living arrangements, beyond keeping a certain distance from a lot of so-called cutting-edge technology and trying to practice traditional virtues of thrift.

Still, basic questions are impossible to avoid. How much time do industrial societies have left, and what is their collapse going to look like? For some people in the most vulnerable locations, collapse has already begun. Is collapse coming for us all? The questions are now common enough to warrant a feature in the *New York Times Magazine* that summarizes the scholarly research on collapse.[19] The same newspaper ran a story on the popularity of a British professor's paper arguing that it is too late to prevent a breakdown in modern civilization in most countries within our lifetimes.[20] Curiously, the article ran in the "Style" section.[21]

Scholars study past collapses and look for patterns that can help us plan for the future, but they have yet to come up with a widely accepted definition of collapse. If one key marker is a reduction in population, what level of human die-off over what time period constitutes a collapse? To what extent do the political institutions of a society have to fall apart to warrant use of the term? From whose point of view do we say a society collapsed? That last question is particularly important, reminding us that not everyone in a society is equally affected by social and ecological disintegration. Vulnerable people may suffer more during the process, but those who are most exploited by centralized power might benefit in the long run from the collapse of that power.

Our goal is not to weigh in on these definitional debates. They are intellectually interesting but not necessary to resolve for our purposes. Instead, we press two simple points: first, we should expect dramatic change in the high-energy/high-technology system that now dominates everywhere, even if not everyone has equal access to the fruits of that system; and second, that change is likely in a matter of decades rather than in some far-off science fiction future. Big change is coming, sooner than any society is ready for.

For our purposes, Jared Diamond's definition of collapse is an adequate starting point: "a drastic decrease in human population size and or political/economic/social complexity, over a considerable area, for an extended time."[22] But we take issue with the subtitle of Diamond's book *Collapse: How Societies Choose to Fail or Succeed*, with its emphasis on choices. One need not deny human agency to realize that once certain forces are set in motion, human choices are constrained by the logic of systems in place.[23] In other words, collapse is not about the failure of a leader or of elites more generally, though when societies fall apart leaders and elites often make bad decisions. Neither is it simply the result of a society exhausting natural resources and polluting the environment, though the degradation of ecosystems is usually part of the process of collapse. Instead, the kind of collapse we're interested in exploring is the consequence of a certain logic in complex systems.

Joseph Tainter's book *The Collapse of Complex Societies* offers a compelling explanation of that logic.[24] A society's leaders invest in complexity to solve what they perceive to be problems. Greater complexity leads to an increase in specialized roles (bureaucrats, soldiers, intellectuals of various kinds) and a corresponding increase in institutional mechanisms to coordinate that activity. Increasing complexity works until it doesn't work any longer; there are diminishing marginal returns to social complexity, a problem that is captured in the idea of needing to run faster to stay in the same place. As societies expand, the control mechanisms (private and public bureaucracies, police, and armies) expand to solve specific problems that arise, providing some cohesion while imposing most of the burden on the least powerful. But as these institutions grow, they become less efficient, and eventually the costs of maintaining the institutions outweigh any assessment of the benefits. That's when a society can no longer solve what it perceives as its problems, leading to decline and—if significant reductions in complexity aren't initiated—collapse.

What does that process look like? Tainter offers a list of features of a society that can be said to be collapsing:

- a lower degree of stratification and social differentiation;
- less economic and occupational specialization, of individuals, groups, and territories;
- less centralized control; that is, less regulation and integration of diverse economic and political groups by elites;
- less behavioral control and regimentation;

- less investment in the epiphenomena of complexity, those elements that define the concept of "civilization": monumental architecture, artistic and literary achievements, and the like;
- less flow of information between individuals, between political and economic groups, and between a center and its periphery;
- less sharing, trading, and redistribution of resources;
- less overall coordination and organization of individuals and groups;
- a smaller territory integrated within a single political unit.

In short: "A society has collapsed when it displays a rapid, significant loss of an established level of sociopolitical complexity."[25] In reviewing that list, we're reminded that collapse is not all bad, that there are positive consequences to the end of complex systems. But our focus is on how to minimize the human suffering and ecological damage done as a system collapses.

There's no algorithm that can predict when the structural trends that create these conditions will result in collapse. Triggering events are unpredictable, and the speed with which systems collapse will vary. But when that process begins, we can expect a loss of "social resilience," the capacity of a society to cooperate effectively to achieve shared goals. Peter Turchin, another prominent scholar of collapse, suggests that structural trends that undermine social resilience have been building in the United States for decades.[26]

How should a society respond? The sociologist Miguel Centeno's focus on the difference between robustness and resilience is helpful.[27] A robust system can endure shocks and remain intact. A resilient system can survive shocks and adapt. In Centeno's terms, robustness is the ability to take a lot of hits before falling, and resilience is the ability to get up quickly after taking the hits. The two goals can be in tension. To build up the robustness of the existing system to withstand the forces of collapse may well diminish resilience in the long term. For example, as ecological degradation makes it more difficult to grow food, we can increase the short-term robustness of agriculture by developing more energy- and technology-dependent methods for keeping yields high to feed the existing population. That will reduce long-term resilience, which requires expanding the use of agroecological methods to produce food in a post–fossil fuel world.[28] Robustness can produce a temporarily stronger but ultimately more fragile system, while resilience increases flexibility and capacity for adaptation. Taking seriously the questions about size, scale, scope, and speed posed in chapter 2, attempts to make the existing system more robust are likely a losing proposition. Better to invest in resilience.

Our preference for resilience is based not only on the high ecological costs of existing systems but also in a judgment that those systems are often an impediment to human flourishing. Take the transportation system, which in the United States has long focused on individual car travel. Investing in electric vehicles may make the existing system more robust in the short term—we may be able to keep driving longer—but that investment in new technology to keep alive an old idea of personal car ownership will undermine resilience. Investing in mass transit, along with the recognition that in a low-energy future we will travel less, will enhance resilience. The ability to jump into a personal car is convenient. But the ecological costs—not only in direct fuel consumption but also in metals and other resources to produce the cars and their batteries, along with infrastructure construction and maintenance—will be just as unsustainable for electric vehicles as petroleum-fueled cars.[29] And along with that calculation we should consider whether a car-based culture that sells us easy mobility has really enhanced human flourishing. We're reminded of Lao Tzu's endorsement of simple living arrangements that leave people so content that they do not yearn to travel. Though they live close enough to hear the cocks crowing and the dogs barking in a nearby village, they never visit.[30] Modern people likely would label that parochial, but the human future is almost certain to be more parochial that the cosmopolitan present.

Preparing, not Prepping or Predicting

One reaction to the possibility/inevitability of collapse is to join the "prepper" or "survivalist" movements—folks who are actively assembling the means and materials they believe will allow them to survive the societal disintegration they believe is imminent. Again, that's not us.

Depending on one's wealth, prepper actions range from ordinary people stocking up on survival supplies and packing "bug-out bags" to the more affluent folks buying space in a survivalist bunker compound to the mega-rich building a community that can float in the ocean and claim political independence.[31] On the truly crazy end of the spectrum is yearning for—and actually spending money based on a belief that someday there will be—human communities on Mars. It's easy to caricature preppers, especially those with reactionary politics who seem to lean toward paranoia and the billionaires with delusions of grandeur. But it's also easy to understand such responses, given the dominant culture's state of denial about the ecological crises.

We are not predicting when a hard-to-define state of collapse will arrive but rather suggesting that change on a scale that merits the term "collapse" is coming and we should be thinking about how to prepare *as a society*. There is nothing wrong with individuals or communities taking action that might enhance short-term survival in crises. For example, it's certainly a good thing for individuals to expand basic skills in food production and storage that will be needed when cheap energy is no longer plentiful and reliable. Being more self-reliant is a good thing. Developing such skills in cooperation with neighbors is an even better thing, enhancing the self-reliance of a group.[32] But those activities should go forward without illusions that individuals or small communities can successfully cope with collapse on their own. No person or community can build a wall high enough or dig a moat wide enough to guarantee survival. And if that were possible, what kind of survival would it be?

Along with any individual and community action, a larger political process is necessary to deal with the dramatic changes coming. Being ready for a radically different life for everyone as part of a radically different ecosphere requires planning. Such a process will need to not only build new political and economic systems but also cultivate a more ecological vision to replace the dominant culture's current linking of a good life to an industrial worldview, what in other writing we have called a "creaturely worldview."[33]

Endeavors at either level—community or country—remind us of the need to revisit what is perhaps the oldest philosophical question: What does it mean to be human? We should expect different people, depending on their talents and temperaments, to focus on different aspects of the challenge. We aren't proposing a specific plan for life on the other side of a collapse. We are suggesting it is reckless not to consider the question. While the two of us live in the United States, the necessary conversation should be international, to whatever degree that is possible. Global economic integration means that while not all people and places are affected in the same way, the consequences of contemporary collapse will not be confined to any one region.

If we were to sum up what's ahead of us in a single phrase, it might be "the old future is gone." We borrow that phrase from the singer-songwriter John Gorka: "The old future's gone / We can't get to there from here."[34] For all of our adult lives, the two of us have lived in a culture that assumed expansion—more people, more energy, more technology, more abundance. Hunger and poverty were problems that could be solved in an ever-expanding world. That was the future we assumed was coming. It's time to retire that notion. Gorka has it right:

> The old future's dead and gone
> Never to return
> There's a new way through the hills ahead
> This one we'll have to earn

There's no point in imagining a world of endless expansion. The new future—the future we are going to have to cope with—will be defined not by expansion but by contraction. It is to that future we have to earn that we now turn.

FOUR

A SAVING REMNANT

In recent years, Jackson has joked that it's time to start a "Center for *Post*-Apocalyptic Studies." If we accept the high probability of coming changes that would warrant the term "collapse" on a global scale, we should be thinking about what lies beyond. What comes after existing social, political, and economic systems are no longer functional? Such a center would concern itself with the idea of "a saving remnant," the human population that would endure on the other side of the collapse.

Jackson was joking and has no plan to launch such a center, though folks at Heidelberg University did just that in 2021, establishing the Käte Hamburger Centre for Apocalyptic and Post-Apocalyptic Studies.[1] The world probably doesn't need another Center for the Study of Anything, but we want to continue to press this point: even if no significant change in the human trajectory toward collapse is likely, efforts to make a softer landing on the other side of collapse are significant. We should be actively pursuing whatever changes we can make now that will make possible that saving remnant.

This approach is not nihilistic. We're not saying that nothing can be done but rather suggesting that there are countless useful things we can do even if we accept the likelihood of collapse. Many people want to do what's possible to prevent, or at least to mitigate the worst effects of, a collapse. We aren't picking a fight with those folks. But refusing to consider the inevitability of collapse can push to the side equally important—and what may turn out to be much more important—planning for the long haul.

This chapter will resonate with those whose thoughts move in that direction. We know there are lots of people who believe that today's multiple cascading crises are manageable within current human systems and that we will invent our way out of trouble. By this time, people counting on technology to come to the rescue will have either closed this book or continued reading primarily for amusement, to see what kind of crazy thing comes next. We don't mind if such people think we're crazy. This chapter is written for people who accept that the old future is gone, struggle to understand what the new future will look like, and ponder what we should be doing now to get ready for it.

We assume that the first instinct of people still reading this book is not to stockpile dried food and shotgun shells, an approach that says, "The hell with everyone except me and my loved ones." There may come a time, if things get desperate enough, when that will be the default response of most people. Such is the premise of many a dystopian work of speculative fiction, which most of us find disturbing precisely because such scenarios aren't hard to imagine. But we're not there yet. Right?

We try to understand what lies ahead and evaluate the most sensible ways to respond. Based on the deep and original insights that we've had into the human condition, we have determined that there are five essential things everyone needs to do, starting right now.

Just kidding.

We have made it clear that we don't play the prediction game, and we also are not in the prescription business. We don't pretend we can see the future, and we also aren't arrogant enough to think we know enough to dictate to people exactly what they should be doing in response to the maddeningly complex challenges before us. It's not just that we have humility about our own capacity to know what's best. Even if we thought we had all the answers, it wouldn't matter. People—and we mean everyone, ourselves included—are only capable of doing what we are capable of doing. That's tautological, of course, but it's worth remembering. We should challenge ourselves, individually and collectively, to expand what we are capable of doing. After all, people don't understand their own capacities with complete accuracy, and those capacities are not static. But we recognize that telling people they must do something that they don't believe themselves capable of doing at a given point in time isn't terribly helpful. More useful is the presentation of a case for what actions are necessary, with the goal of helping one another get to the point where such action is possible.

We have already recognized that people have widely varying talents and temperaments, which lend themselves to some pursuits more than others. No matter how much we might be able to get out of our comfort zone, we still tend to act within our talent-and-temperament zone unless faced with an immediate threat that forces us out. We—and, again, we include ourselves in this analysis—often tell ourselves a story that the right thing to do just happens to coincide with what we feel ready to do or already wanted to do. Rather than lecture others about what they should do, we're all better off struggling together to understand why difficult choices that we may be reluctant to make are necessary. Rather than harangue others about why they should adopt our analysis—which, of course, we believe people should adopt, or else we wouldn't be writing this book—we explore some questions that are of value no matter what projects people might be inclined to pursue.

When people say, emphatically, "We have to do something now!," we agree. But what we decide to do now will depend on our analysis of what is possible and most effective. Reasonable people can disagree about that and pursue different projects motivated by the same values. For those of us who want to face a new future, what are our options for action? For purposes of clarifying our choices, we suggest three broad categories:

1. A focus on the present. Change society in the realm of both public policy and everyday cultural practices in whatever ways are possible right now, not because we can avoid collapse, but rather to minimize the human suffering and ecological damage we do on the way there.
2. A focus on the coming years. Accept that collapse is coming and prepare local communities for the transition as society is forced to

adapt to new realities, anticipating to the extent possible the conditions under which that transition will happen and hoping it will be orderly enough that we can hold onto our humanity.

3. A focus on what lies beyond. Accept that collapse is coming, recognize that the conditions under which the transition will take place are likely to make an orderly process difficult if not impossible, and focus on what humanity will require in a future defined by "fewer and less."

These three categories are not mutually exclusive. At any given moment, it is not obvious what will be the most effective actions. In our own lives, the two of us have at various times pursued projects at all three of these levels, many of which we believe have been productive, generating positive outcomes that we either hoped for or were pleasantly surprised by. But today we feel the third focus does not receive adequate attention, so we have decided to train more of our attention there—what we refer to as the "saving remnant" option.

Biblical Roots of the Concept "Remnant"

The most common use of "remnant" is to describe a small portion left over, what remains of a piece of cloth or a roll of carpet that is too small to be used for the job at hand. These remnants usually end up in a drawer or a corner of the garage until a new use can be found for them, perhaps stitching them together for a new project or using them as a patch. We borrow the phrase "saving remnant" from Jewish and Christian traditions. (As is always the case when we cite religious traditions and scripture in this book, we are not making theological claims but using religious language to help us understand human struggles beyond the sectarian.)

One theologian defines the saving remnant as "what is left of a community after it undergoes a catastrophe."[2] The account of the Flood (Genesis 6:9–9:17) is in some sense the foundational story of a saving remnant: Noah and his family and the animals on their boat. The concept is invoked explicitly by Hebrew Bible prophets such as Isaiah

(Isaiah 1:1–9). The term is used in various ways, but at the core is a faith that even in the face of an overwhelming catastrophe, a saving remnant will survive and become the basis for renewed community life. In some places in scripture, a remnant survives even though everyone deserved destruction. In other places, the survivors were the righteous and faithful. As one biblical scholar put it, "However great Israel's apostasy and God's judgment, a core of the faithful will still exist."[3] But even with that promise of renewed life, people such as Isaiah made it clear that God was never happy with Israel's failure to honor the covenant and create a just society.

> In that day the remnant of Israel, the survivors of Jacob, will no longer rely on him who struck them down but will truly rely on the Lord, the Holy One of Israel.
> A remnant will return, a remnant of Jacob will return to the Mighty God.
> Though your people be like the sand by the sea, Israel, only a remnant will return. Destruction has been decreed, overwhelming and righteous.
> The Lord, the Lord Almighty, will carry out the destruction decreed upon the whole land.
>
> (Isaiah 10:20–23 NIV)

That's a reminder that even if you are part of the saving remnant, it's best not to get cocky, because there is a more powerful force (God in Isaiah's case or the ecosphere in our analysis) waiting to humble you.

This idea of a remnant shifts in the New Testament, since Christians—many of whom were faithful Jews—were arguing that there was a new covenant in Christ. For Paul, that meant God was not giving up on Israel but offering a new arrangement, salvation by grace through faith (more on a secularized concept of grace in chapter 5).

In more recent history, the concept of a saving remnant has been invoked in a variety of religious and secular contexts. During the First Great Awakening in eighteenth-century New England, some evangelical Protestant clergy, who were called "New Lights" and supported revivals, compared themselves to the saving remnant of scripture.[4] The concept endured in Judaism as well. Doctors in the Warsaw Ghetto adopted the motto, "Non omnis moriar," which can be translated as "I shall not wholly die" or "Not all of me will die." Those doctors had faith that "some Jews will survive somewhere, and some fragment of civilization will survive which those Jews must help to refresh. Though all friends, all relatives, all reasonable hope may perish, the Saving Remnant will not die."[5]

W. E. B. Du Bois invoked the term in his famous 1903 essay on African American leadership, "The Talented Tenth." White society had "lynched Negroes who dared to be brave, raped black women who dared to be virtuous, crushed dark-hued youth who dared to be ambitious, and encouraged and made to flourish servility and lewdness and apathy," Du Bois wrote,[6] but it could not destroy black people's aspirations.

> A saving remnant continually survives and persists, continually aspires, continually shows itself in thrift and ability and character. Exceptional it is to be sure, but this is its chiefest promise; it shows the capability of Negro blood, the promise of black men.[7]

A secular leftist historian in the United States used the phrase to describe the relatively small number of activists and organizers who are "neither indoctrinated nor frightened into accepting oppressive social conditions" and openly challenge power, striving "to awaken and mobilize others to join in the struggle for a more benevolent, egalitarian society."[8]

To be clear: For our purposes, we use the term without speculating on divine judgment or weighing in on which group has God's favor and without suggesting that any particular political ideology is required to be special. We aren't fond of sectarian struggles and try to avoid the temptation to condemn those less righteous than we believe ourselves to be. It's helpful to recall Heschel's wisdom cited earlier: "Few are guilty, but all are responsible." Or if one is more comfortable with a secular version, there's Walt Kelly's famous 1970 Earth Day poster and 1971 cartoon, in which Pogo says, "We have met the enemy and he is us."[9]

It's easy for people who believe themselves to be among the chosen to declare themselves the rightful keepers of the covenant, the ones who deserve to be part of the saving remnant. Given our ages, the two of us don't expect to be around long enough for that next stage, and if we were younger we wouldn't be claiming that we deserve to be among that group. For us, the attraction of the idea of a saving remnant isn't about righteousness or superior ability but rather how to think about the softest landing possible for what will be left of human society in a difficult future.

What will be required of people in that uncertain future? What will life on the other side of today's high-energy/high-technology cornucopia look like? No one can predict the terrain, social or ecological, on which people will be building a different world. But we expect it to be defined by "fewer and less"—fewer people consuming less energy and resources. Whatever the specifics of that future, we can contribute today by thinking about the *stories*, *skills*, and *spaces* most needed. But before thinking about a low-energy future, we should reflect on our living arrangements in the existing high-energy world and get clear on the problems those arrangements create for our ability to imagine a new future.

Wants and Needs

People with a similar view of the human trajectory and similar objectives can pursue very different projects. That's fine with us. If no one can know for sure how to achieve common objectives, as Mao Zedong said, "Let a hundred flowers bloom, and a hundred schools of thought contend." Reasonable people can and will disagree on priorities. But while people may choose different actions, we can stay connected and struggle together with difficult questions, especially those foundational questions that are relevant to all our projects.

Here's one of those questions. How do we decide what we really need? How do we distinguish those needs from things we have come to want? The question matters, because as the anthropologist Marshall Sahlins puts it, "There are two possible courses to affluence. Wants may be 'easily satisfied' either by producing much or desiring little."[10] Since we cannot produce much indefinitely, desiring little is going to become more important.

Those questions matter for those of us who believe we are heading toward a fewer-and-less future—no matter whether our work today focuses on the short-, medium-, or long-term responses to the multiple cascading crises. It's uncomfortable to ponder how we can distinguish our core needs from mere wants that we can live without. People—again, we include ourselves in this assessment—go to great lengths to avoid the question because it forces us to confront our own weaknesses and failures, our own tendency to self-indulgence and frugality avoidance.

We know that the most basic needs for survival are food, water, clothing, shelter, healing, security, and sociality. We need enough to eat and drink, the means to avoid life-threatening or extremely uncomfortable elements, ways to heal from routine ailments, and protection against predators, human and nonhuman. Because we are social animals, intimacy and membership in a community also are basic needs.

Everything else in our lives is in the category of a want, a desire, a yearning—the things we can live without and still live fulfilling lives. If any one of us living in the affluent world were to make a list of what we consume in a given day—all those goods and services that can be so easily purchased in an affluent society—the list of wants we seek to satisfy would be much longer than the list of needs we fill. The only people for whom this isn't true are those so poor that they struggle to meet those basic needs and an occasional spiritual ascetic or off-the-grid hermit.

Here's a trivial example. We both start our morning with coffee. We are tempted to say that we both need coffee to get going in the morning, but of course coffee is not really a need. Lots of people live without coffee—either by choice or because of circumstance—and get along just fine. Coffee is a want, something on which we have become dependent but something we can live without. How did we become so hooked? How much of it was a real choice, and how much was a response to cultural norms we never thought to question or to advertising and marketing that led us to this dependency?

Those are hard questions to answer. But here's an easy observation. In a fewer-and-less future, people living on the Kansas prairie and in the mountains of northern New Mexico will not be drinking coffee unless we're willing to sacrifice a lot of other things to get coffee beans that don't grow where we live. For us, coffee is a great example of what Wallace Stegner described as "things that once possessed could not be done without."[11] There are many things that we believe we can't do without today that we will do without in the future.

This is a trivial example because coffee-or-no-coffee isn't really going to drive anyone to despair. But, trivial as it may be, we don't want to think about life without morning coffee. We don't want to ponder how we came to think of coffee as a need, not so much because that morning routine is so important, but because it raises troubling questions: How many of our perceived needs are really wants? What are we willing to do about it? Why are we so hesitant to talk about these choices?

Here's our gut feeling. As a society, we find it hard to talk about needs versus wants partly because of the length of the list of things we feel that we need but that we know are mere wants. But just as scary is the awareness that in a future with more dramatic limits, we are going to have to sort all this out together. We all should reflect on our conversion of wants into needs, but when resources are limited that will have to be a collective conversation. Right now, as long as people have the money, they have to justify their needs/wants calculus only to themselves or, perhaps, family and close friends. Even that can be difficult, especially if family budgeting is tight. How do we evaluate one person's desire for just the right beans to make morning coffee against another's desire for an overpriced scone from the bakery down the street? Again, this may seem trivial, but in a society in which many people's sense of identity is so often connected to lifestyles and products, what happens when we no longer have the resources for most of it? Whose wants, desires, and yearnings will count, and whose will be ignored? How do we resolve the differences of opinion after so many years of indulgence?

This process of down-powering should not be confused with the "new minimalism" fad of the 2000s, the project of "decluttering and design for sustainable, intentional living."[12] The use of "sustainable" in that context has little or nothing to do with ecological sustainability. There's nothing wrong with "tidying up" one's home or office,[13] and it may improve one's mental health. But we shouldn't expect such superficial changes to be "life changing,"[14] if by that we mean a real shift that slows or reverses the life-destroying consequences of the high-energy/high-technology world. There is now a small industry that helps affluent people continue to live affluently but with less of a mess. We don't find that a source of inspiration.

Other approaches to simplifying our lives have included an awareness of the ecological crises and, hence, are more useful. The frugality and simplicity in Warren Johnson's *Muddling toward Frugality*,[15] first published in 1979, and Duane Elgin's *Voluntary Simplicity*,[16] first published in 1981, were grounded in the larger analysis of human overshoot circulating at that time. But perhaps even the current bourgeois bent of consumerist minimalism is better than nothing. The journalist Jia Tolentino suggests that the most compelling case to be made for the fad is what it could help us confront: "With less noise in our heads, we might hear the emergency sirens more clearly. If we put down some baggage, we might move more swiftly. We might address the frantic, frightening, intensifying conditions that have prompted us to think of minimalism as an attractive escape."[17]

Perhaps Tolentino's conclusion is warranted, but we're focusing on the day when this move to simpler living will not be voluntary. Frugality will have to be imposed through collective action. Whether that imposition will involve what we recognize today as "government" or some new form of political association depends on a trajectory we cannot predict. But at some point, "fewer and less" will not be a matter of choice but will be reality, whether we like it or not. We will have to manage collectively the choices within new limits. We see no reason to believe that, in the time frame available, human societies will embrace public policies that significantly change the collapse trajectory. But once we are faced with new limits, societies will need to work out strategies for democratic self-management of resources at the local level, allowing people to hold each other accountable without centralized power. How to do that is neither easy nor obvious.

We need to begin difficult conversations about needs and wants to make that hopeful future more likely (more on hope in the conclusions). Perhaps such conversations should not be difficult, but they are for a lot of people. A capitalist culture dependent on mass consumption cultivates a close connection between people and our stuff. Many people's identity is tied to activities that require buying a lot of goods and services, and we should not be naive about how much struggle will be involved in leaving that consumption behind. But any project to make possible a saving remnant should start by acknowledging that most of our stuff—including the stuff that we think is important in defining who we are—is not only ephemeral, but soon will be unavailable.

We don't just need to stop consuming so much. We need to create new ways of understanding ourselves when consuming isn't such a big part of our lives. We need to tell new stories about ourselves, learn and teach skills for new living arrangements, and create new spaces for this work. But before discussing those endeavors, we want to linger a bit longer on the discomfort many of us feel about our consumption.

A New Identity Politics

There is a spirited debate among US progressives about "identity politics." That term describes the project of developing and advancing political agendas from the shared experiences of injustice based on such categories as sex, sexual orientation, race, and ethnicity—or some combination of those identities. Sometimes this approach is in tension with a class-based politics that puts economics and the control of wealth at the center of the political struggle. We need not weigh in on these fraught debates (and if we did, we would take a cautious position that would highlight the complexity, seeing some merit in many of the arguments) to point out the centrality of another identity in most of the world: Consumer.

People have always consumed things, of course—food, water, and various resources, those naturally occurring as well as manufactured by people using tools—but it is only recently that capitalism created the widespread identity of the Consumer. Our analysis of twenty-first-century identity politics starts with a simple claim: there are many identities that are the basis for differences among us that require continued attention in the pursuit of justice. And at the same time, we need to renounce the common identity of the Consumer. We need to tell stories about what it means to be human beings who consume to satisfy needs rather than people who conform to capitalism's illusions to satisfy wants.

Our rallying cry: The end of the Consumer in us all!

We may not want to see ourselves this way. Indeed, both of us bristle at the thought of being lumped into the Consumer identity, since we pride ourselves on not keeping up with fashion or every new gadget. But it would be folly not to recognize and try to root out the Consumer in ourselves.

We don't confine this identity to those who clamor for consumer goods that no one would consider necessities. Consider the people who line up to buy the latest smart phone or those who eagerly await the new version of a video game. Folks who shell out lots of money to attend spectator sports and buy team jerseys are Consumers. So are people who flood into a stadium to see a pop music star and buy the concert T-shirt. And the people who open their wallets for premium coffee and trendy baked goods. And the people who special order granite countertops to ensure the ultimate cooking experience. And on and on, up and down the aisles of stores, click after click on the websites.

We're not saying that communication devices, home entertainment, sporting events, concerts, caffeinated beverages, breakfast pastries, or well-built kitchens have no value. We are saying that in a mass-marketed world, even items of real value get sold to us in ways that are as much about how we feel as what we need. Jensen winces at the thought of the many years he bought books that he knew he likely would never read, simply because seeing those volumes on his bookshelf made him feel good and, he hoped, would convince others that he was smarter than he was. Jackson can point to the four vehicles in the driveway of his two-person home (older vehicles but one car and one pickup per person), which he might argue is at least two and maybe three vehicles too many.

If any readers protest and claim to have rejected the "Consumer" label for themselves, that's fine. But whatever self-image we have, the advertising and marketing wings of capitalist enterprises are clear about how they see us. "The way consumers see themselves determines their behavior—and you can influence that," promise three business professors writing in the prestigious *Harvard Business Review*. "It's surprisingly easy to make consumers switch identities and even to give them new ones."[18] The business literature is crammed with research on and strategies for how corporations can create and manipulate our identities and spending habits, mostly in the context of getting us to buy specific goods and services. But our focus is not on businesses' competition for our allegiance to specific brands but rather the allegiance to consuming, the way a consumer capitalist economy needs us to embrace the identity of Consumer.

Rejecting the Consumer in us all and embracing a new identity isn't so easy. Capitalism has depended on both state and private violence to get established and survive; think of imperial wars to acquire resources and establish markets and decades of strike-breaking. But the strength of the identity of Consumer is maintained primarily through pleasure. For most people it feels good, at least in the short term, to consume. Once we settle into the Consumer identity, it's surprisingly hard to escape, especially given how mass media and social media reinforce the pleasures of consuming as our capacity to imagine alternatives atrophies. As one critic puts it, the problem is the limits of our imaginations.[19] We need to expand our imaginations, which requires different stories.

New Stories

A saving remnant will need new stories born of resistance to the Consumer. What stories can we tell about what it means to be human that will help us on the other side of collapse? The values in many of humanity's stories that come from a time before the Consumer will be helpful, but the circumstances humanity faces today are unprecedented. There have been societies that suffered catastrophes in the past, of course, but those typically are stories of a forced contraction, often involving violence that was used to subdue or destroy a culture. As one Indigenous writer-activist put it, "To be Indigenous to North America is to be part of a postapocalyptic community and experience."[20]

The possibility of that level of barbarism within the human family still exists, but the challenge today is the human family's inability to face collectively the need to move to a world of fewer-and-less. We need stories not just about how to live in a low-energy world, but stories to help us negotiate the unprecedented journey that humanity must make from a high-energy to a low-energy world. As we keep emphasizing, this is a new challenge. No culture, past or present, has faced these circumstances. There is no template.

Stories work best when they are familiar. Some old stories may be useful, but they were created in another time for another world and will need modification or wholesale change. We especially will need a story about why so many people continued to be so short-sighted and cruel, even when the knowledge of coming collapse is widespread. That should lead us to keep telling, or resurrect, old stories with themes of justice and sustainability. But coping with a global collapse is unprecedented, and we cannot expect old stories to be adequate. We cannot expect to return to some previous state of being. We are going to have to craft stories about a new place in the ecosphere. The values of old stories will be important, but the values they exhibit have to be lived in new circumstances. Another Indigenous writer describes what lies ahead as "going through the narrows," which requires us to recognize the difference between wants and needs. "Life will be difficult, at best, for some, and unbearable for others—and costly in many ways for everyone."[21]

Neither of us has the kind of creative capacities necessary to tell a new origin story or make new myths. But we understand their power, which is captured in the observation of an eighteenth-century Scottish politician: "If a man were permitted to make all the ballads he need not care who should make the laws of a nation."[22] The most efficient, and perhaps effective, way to control people is not through the coercion of law but through the power of stories. The most effective way to inspire people is with stories. We aren't arguing for art-as-propaganda but recognizing that much contemporary storytelling serves a propaganda function for the state and corporations. The new stories shouldn't be instruments of control by a few but a collective act of expressing inspiration for everyone.

It's easy to identify some of the ballads that are not helpful. Stories that glorify military conquest, celebrate national greatness, or assert claims of inherent cultural superiority are destructive. Such stories encourage the illusion of human control, over others and over the larger living world. Equally unhelpful are stories that focus our attention on media celebrities, spectacle sporting events, the thrill of gambling, luxury vacations, designer clothing brands. Stories that valorize rich people are a real problem. These kinds of stories numb people in the face of painful realities and foster illusions about what is of real value. We have a hunch that the most powerful stories won't be about extraordinary individuals but about regular folks who persevere through difficult times and take care of each other.

It's also easy to observe that in the affluent world we have turned over most of the storytelling to professionals, and we too often are reduced to being Consumers of the creativity of others that is packaged and sold to us. We're grateful for the creative work of those writers, musicians, and filmmakers, some of whom are our friends and, for one of us, a spouse. Such artists have produced work that has enriched our lives, and some have already begun the work of speaking apocalyptically and telling new stories.[23] But in a saving remnant future, ordinary people will have to do more of the things that in a high-energy and affluent world are turned over to professionals. One of the many skills we need to rebuild is the deeply human ability—an ability everyone has—to tell stories.

Skills

Other skills will be needed to live in a low-energy world. As the industrial world runs down, there will be a transition period in which we will need lots of people to patch together the infrastructure high-energy societies created. Given the complexity of that infrastructure, we will have to assign more value—not just in terms of compensation, but more important in status and prestige—to the specialists who either built or maintain those systems. Electricians, carpenters, construction workers, plumbers, and people with applied engineering skills of all kinds will be the truly essential workers in a down-powering world.

Today those jobs may pay well, especially for those workers represented by a union, but they are rarely seen as high status. That's in part because they do not require a college degree, the current stamp of social status not only in the United States, but in much of the world. There's no need to denigrate intellectual workers, but in hard times it is hands-on skills and experiential knowledge that carry the day over management skills and abstract understanding. We don't need to abandon intellectual work but rather increase the respect we accord to labor that has for too long been misunderstood and underappreciated.

The same applies to agriculture and rural life more generally. In a low-energy future, agriculture that is not so heavily dependent on fossil fuels and does not answer to agribusiness multinationals will require more people on the land with more practical skills. Human and animal muscle power will replace the dense energy of oil and natural gas that currently provides traction, fertility, and weed and pest control. Much farm labor is hard work and also requires complex knowledge and skills that precious few in the United States and other affluent countries still have. Along with repopulating the countryside, we will have to collectively relearn how to grow food without cheap energy. Ironically, those people living farthest from the temptations of dense energy—what many affluent people look down on as "the peasantry"—possess far more of those skills than the affluent.

A low-energy future will require everyone to be skilled at living with less, including those who are wealthy, since current methods of hoarding wealth in financial instruments eventually will become irrelevant. Many people are already moving in this direction, learning to garden, hunt, preserve food, and cook without prepared foods. And while skilled tradespeople will be essential for dealing with complex infrastructure, everyone will need to be capable of "tinkering," repairing things in our domestic and work lives that are designed to be thrown away and replaced with new items.

Beyond the specific skills, what is most important to an ongoing human future is a belief in our ability to survive without all that energy. This is not about a nostalgic return to a simpler world; because of the high-energy infrastructure all around us, nothing will be simple. Instead, it is about the transition from high to low energy. If we are going to pull that off, we have to believe we can pull it off. That means we need to demonstrate to ourselves that we are competent to manage our own lives, and anything that allows us to do that is positive. Many of us have experienced the satisfaction of learning new skills, evidenced by the number of YouTube videos that explain how to do, for example, basic home repair. Online resources are incredibly helpful today, but we should not neglect the importance of face-to-face communication, aware that in a low-energy future that will again become the dominant—and perhaps eventually the only—form of human interaction.

Spaces

One of the most jealously guarded assets in the modern world is privacy. Wealth can buy distance from other people, in both social and geographic terms. Rich people can buy whatever they need without having to mingle with the public and can live in big houses on big estates that keep people away. As the digital world shrinks privacy in other ways, spatial privacy has become even more valuable to many.

In a down-powering world, such privacy will not only be more difficult to obtain; it will be less attractive. We will need more public spaces for the kind of human interaction that doesn't depend on high-energy/high-technology communication infrastructure. The construction of such spaces requires rethinking contemporary living arrangements in affluent cultures, from the obsession so many have with spacious private homes to the layout of towns and cities. It also requires a shift in our thinking about human relationships. Instead of constructing our lives around protection of the private sphere (typically family and close friends) we will have to learn to anchor our lives in the public sphere (typically the concept of community). Some people obviously already live by these values, which will have to become the norm throughout the culture.

What spaces can we create to foster the human interaction needed to sustain a new human community? To imagine spaces that will exist when there is no widely accessible internet, we are going to have to spend more time in face-to-face engagement. If that doesn't happen through routine interactions, how do we create new routines?

Throughout history in different societies, the layout and use of public space has varied depending on culture and geography. For purposes of imagining a low-energy future, we pay attention to the lessons learned from religious gathering places: a church, synagogue, temple, or mosque. Again, we aren't advocating for any particular religion, and we don't build our own worldviews around any traditional theology. We are thinking about what keeps people connected, and religion is certainly a powerful connector. Most debates about religion involve doctrine and dogma, what people believe and how they worship. We want to put such wrangling aside for the moment and focus on what religions offer people in addition to belief systems: a physical space and a social routine. As we think about a shift to more community-based life in a low-energy future, these things seem more important than ever.

We use the term "physical space" instead of "building" to recognize that while most religious communities come together in a building dedicated for that purpose, it's not the structure that matters. What's important is having a place where everyone knows to gather, whether it's a fairly permanent temple or a deliberately temporary tabernacle. We are convinced that a big part of the attraction of religion—so important that many people keep coming long after they no longer believe in the doctrine or dogma—is that sense of familiarity and comfort in a familiar space.

Just as important is the routine of worship, the structure of a service from opening prayer through scripture readings and hymns, followed by a sermon, with more singing and final prayers (the typical experience in Protestant churches, in which we grew up). Of course, people listen to the words and ponder their meaning, but the familiarity provides much of the comfort.

Whatever one thinks about the behavior of religious leaders or the intellectual coherence of any particular theology, we can learn much about how to organize low-energy communities from paying attention to the spaces and routines of religion. We realize there are secular spaces, such as labor union halls or social clubs, that at various times have served the same function. One might be tempted to cite bars and pubs as well. Recall the chorus of the theme song of one of the most popular US television shows of the 1980s and 1990s, *Cheers*, which celebrated places not only where everybody knows our name and greets us warmly but also where we are reminded that "our troubles are all the same."[24]

Belonging to a community-sized congregation feels much the same. There is a fellowship of familiarity, an assumption of acceptance, and shared struggle. But unlike a bar, there's also a commitment to a shared worldview and accompanying ethic, even if it is lived imperfectly. There's a great deal of hypocrisy, as members of a congregation articulate noble ideals but ignore them in practice. We all have the potential to fall short, and at some point we all have. But it's important that the ideals are articulated and regularly reinforced in that space, and returning every day or week to be reminded of those ideals is a good thing.

Milling Around

When one identifies a problem, one can be sure that before too long someone will say, "We all know the problems. What's your solution?" This comment is taken to be a sign of seriousness, that the person is not interested in complaining but in getting things done. But such comments can be an impediment to meaningful conversation about solutions.

First, it's not accurate to say that we all know the problems, even among like-minded people. For example, there are deep divides in the environmental movement over how to understand the problems that emerge from multiple cascading crises.[25] Different analyses lead to different assessments, and those differences are not trivial. Take our previous example of transportation. We believe the problem is a car-based culture, while some environmentalists think the problem is cars that burn petroleum. We think electric vehicles are a diversion, while some others see them as a solution.

Second, sometimes there are no solutions within the existing framework and expectations, and it's important to tell the truth about those limits. As we have made clear, we see no solutions to the problems of social justice and ecological sustainability, if by "solution" people mean eight billion people on the planet and high-energy/high-technology lifestyles for a big chunk of those people.

When we accept that the existing framework cannot endure, then new possibilities open up. Some might say our approach to a saving remnant is no solution at all, but we argue that it's not only viable but inevitable as well. Attempts to keep the existing systems going will simply accelerate the movement toward collapse and leave future generations with fewer options. We believe the saving remnant "solution" is pragmatic.

Still, the typical response to a presentation about a social problem is, "What should we do?" Our approach suggests that those of us living in affluent societies start with a different question: "What are we willing to give up?" The "we" in that sentence is important, though as individuals we should all ask ourselves that question. The path forward requires a collective reckoning with that question.

We have highlighted what we believe are some of the impediments to this work, but we should not forget the assets we have. A quick review of our evolutionary history reminds us that we evolved to take care of each other in small groups that can maintain a rough equality by developing social practices to control human arrogance and greed.[26] We have a cognitive capacity that allows us to make tools that compensate for our relative lack of strength and speed. In the ten thousand years of agriculture and empires, we've made "progress" that leaves us in dire straits. But we can remember that not only is there another way, but that for most of human history our species lived in other ways, and some still do today.

There will no doubt be many variations in how a saving remnant manages to survive on the other side of collapse, depending on geography, climate, and environmental conditions. And because we won't be starting from scratch, the many different cultural histories that people carry into the future will influence outcomes. But there likely will be some common features, captured in one of Jackson's favorite philosophical formulations. For some years now, Jackson has advocated what he calls the "mill around" philosophy of life. "How do we mill around, amuse ourselves, and live cheaply until we die?"[27]

We have emphasized that we shy away from prediction and prescription. We recognize human limits at forecasting the future with precision and are cautious about the arrogance that leads people to tell others how to live. But we offer the "mill around" philosophy as a standard for evaluating the projects we might undertake. The most important experiments in coping with unprecedented challenges are the ones that help us amuse ourselves as we mill around and live cheaply until we die.

There's a saying in the minimalist movement that less is more; a good life is not only possible but also more likely when one reduces consumption. Fair enough, but it's perhaps more accurate to say, "Less is less, but less is OK." Whatever we may want, our future will be marked by less consumption. The pleasures of consuming will have to be offset by other activities that provide satisfaction. The more people there are embracing a mill-around theory of life, the more we will be able to accept less.

We might surprise ourselves and each other by being stronger, braver, kinder, and more compassionate than we thought ourselves capable of when a crisis demands those virtues. But we also should remember that we can be weaker, more cowardly, meaner, and less compassionate than we thought. In crisis times, those failures are also possible. We can hope that the better angels of our nature will emerge, but the devil is in the details.

The sooner we confront those devils, the better.

FIVE

ECOSPHERIC GRACE

By now, readers likely have noticed that for two people who don't belong to any religious organization, we use a lot of religious language. Throughout the book we've borrowed concepts from Jewish and Christian scriptures to frame our arguments. This is a questionable strategy, since it risks offending nonbelievers (who might wonder why we have to be dragging in all this God-talk, even if we leave God out of it), believers from those faiths (who might be critical of how we use the tradition without accepting its doctrine and dogma), and believers from other faiths (who might wonder why we don't draw on other traditions).

The two of us long ago accepted that we are products of a predominantly Christian culture and its stories. We are in some sense deeply Christian and at the same time thoroughly secularized. Because there is no plausible evidence for the tradition's supernatural claims, we don't accept the existence of God as a being, entity, or force that acts in the world; the virgin birth and resurrection as historical events; or any life after death. But we can't shake using some of the concepts of the tradition, especially as we think through social problems.

We aren't just endorsing the easy to endorse moral teachings. We agree, for instance, that we should try to love our neighbors as we love ourselves, but that sentiment is hardly unique to Christianity. We are recognizing that in our formative years, one of the ways we learned to see the world was Christian and Protestant. That framework was part of the culture we were raised in, and it had a role in shaping us, independent of our evaluation of doctrine and dogma. Today, we still like a lot of the stories, not just because they are good stories, but because they help us organize our thoughts about the world. Instead of always fighting our history, we try to work with it.[1]

Here's one of those connections you may have noticed: our discussion of the immense changes that came when humans domesticated plants and animals, which turned us into a species out of context, sounds a lot like a story about "the Fall" and the notion of "original sin." When did things start to go off the rails in a big way for *Homo sapiens*? It was when we humans started believing that we could take control of landscapes and bend other living things to our will, the act of domestication. We assume humans have always had the capacity to be violent and destructive, which no doubt has meant that throughout human history some humans have caused trouble. But the real trouble started when people got the idea that we could own the world, when people started to play God, intervening in ecosystems in ways that would eventually cause lasting damage.

Jackson has for years been retelling the story of Adam and Eve's banishment from the Garden as a cautionary tale about the limits of human knowledge. In chapters 2 and 3 of Genesis, God told the couple that they could eat freely of every tree in the Garden except the Tree of the Knowledge of Good and Evil. Why were they tempted to rebel? Blame it on the 1,300-cubic-centimeter brain (the average size of an adult human brain). That brain gives us the cognitive capacity to do extraordinary things, which leads to extraordinary results, both constructive and destructive. We usually hold up the positive results as evidence of our brilliance, and we push to the side the unintended negative consequences.

That's civilization's original sin: hubris, a lack of humility. We overreach, believing that because we can know a lot we can know enough to play God. The result is millennia of soil erosion and degradation, along with the beginning of a reduction in biodiversity, now intensified and joined by chemical contamination and rapid climate destabilization as threats not just to our "quality of life," but to life itself. Not every person acts with disregard for other life, of course, nor has every culture developed the technology to do that kind of damage. But our species' capacity to undermine the health of ecosystems has led to that result in many times and many places since the invention of agriculture.

We find this reading of the Genesis story as a warning against arrogance instructive and also comforting. It doesn't make our struggles to cope with that Fall any easier, but we take heart in knowing that contemporary humans aren't the first to wrestle with it. It's not just our failure today, but a part of a larger struggle for the past ten thousand years. This kind of wisdom is not found exclusively in the Bible, or in any other religious text, but the Bible and religious texts are one place where it can be found. Like it or not, the Hebrew Bible and the New Testament supply many of the stories that we in the United States, along with much of the rest of the world, were raised with. Those stories aren't going away.

Where does that leave us? Desperately in need of grace.

Christian Grace

In Christianity, "grace" is understood as God's love and mercy for people, even though we have not earned it. Most Christians believe that because God loves us, we can receive a kind of "unmerited favor," as grace is often defined. God's grace is generous, free, undeserved, unexpected. All we have to do is have faith in God. "For it is by grace you have been saved, through faith—and this is not from yourselves, it is the gift of God—not by works, so that no one can boast" (Ephesians 2:8–9 NIV). The author of that passage, the apostle Paul,[2] also encouraged people to do good works, of course, but since we are all sinners—another religious idea that is hard to argue with, since we've all done things we know were wrong and hurt others—we all need grace.

There's one troubling aspect of the idea of God's grace, however. It appears that we're the only organism on the planet that receives such unmerited favor. Scripture doesn't talk about moose or trout or cockroaches getting grace. That would suggest we have, and somehow deserve, favored status, which leads to trouble. Many people want to believe that we humans are the special objects of his attention.[3] God holds us accountable and demands that we have faith beyond evidence and reason. If we are faithful, we get grace because God loves us, and apparently loves us more than he loves the other creatures. While subordinating people to God, this doctrine gives humans superior status to other life on the planet, which turns out to be a very dangerous notion when we take it as license to subordinate other creatures, not only to our needs, but also to our wants. That idea can undermine our humility. After all, if we are so special, it's not surprising that we find it easy to believe we should be running the planet. That idea shows up in other places, starting early in the Bible: "And God said, Let us make man in our image, after our likeness: and let them have dominion over the fish of the sea, and over the fowl of the air, and over the cattle, and over all the earth, and over every creeping thing that creepeth upon the earth" (Genesis 1:26 KJV).

This passage is cited by some Christians as the basis for "dominion theology" or "dominionism," taking it as a biblical mandate for Christians to control all institutions, including secular government. More commonly it is interpreted as a call to stewardship, since one is responsible for what one has dominion over. But stewardship doesn't really solve the problem, since that concept implies control and superiority as well. Even if we try to be better stewards, we are still claiming to be in charge, and to date we have failed in that role.[4] Remember that to acknowledge that a tradition carries wisdom is not to claim that everything in a tradition is particularly wise. This is one of the places where serious rethinking of the Jewish and Christian traditions is needed.

Still, we want to hold onto the concept of grace, the idea that we don't constantly have to earn our standing in the world but can do our best and be secure in our place on Earth. But we want to drop the invitation to arrogance that comes with being special or claiming dominion. Could we replace conventional notions of God's grace with ecospheric grace? This isn't such a crazy idea, given that when some people—maybe even most people—hear the word *God* they think not of a being like us, or any other kind of tangible entity that exists on a purely physical plane. For lots of folks, "God" is a synonym for "nature," for the larger forces that make our lives possible. That's why so many people, including traditional believers, will say they feel closest to God not in church but in places relatively undisturbed by humans—walking through a forest, climbing a mountain, sitting by a stream, swimming in the ocean. We're not talking here about formal theological arguments but rather about what many people report feeling.

God is said to be transcendent and ineffable, something beyond us and beyond our capacity to understand. That is also true of the ecosphere. We understand a lot of specific things about this planet and the life forms on it, but we should recognize that our ignorance is greater than our knowledge. This is especially true when it comes to the complexity of life, the interactions among living things, and the interactions between what we understand as the living and nonliving components of the ecosphere. We cannot come close to naming all the component parts, let alone understand all the interactions. Science allows us to begin that inquiry, but sensible scientists realize how quickly we bump into the limits of our big brain.

Asking people who say they believe in God to define God is a bit unfair, since by definition God is beyond definition. But in practice, many people's understanding of God maps pretty closely onto the ecosphere, to the power of the nonliving world to give rise to life, to the (nontheological) miracle of it all. Hence, ecospheric grace.

Who Loves You?

We like the idea of ecospheric grace because it doesn't depend on the ecosphere loving us or bestowing on us special favor or giving us dominion over anything else. That's important because, as far as we can tell, the ecosphere does not love us. The ecosphere does not care that we exist. We are, in ecospheric terms, just another species in a long list of species that usually end up going extinct at some point.

We may love the ecosphere. We may feel reverence when we have a moment of awareness of the complexity and beauty and power of the larger living world beyond us. The vastness of the universe is a never-ending source of amazement for most of us. We may feel awe when we are fully open to the power of what people routinely call "nature," though we should always remember that humans are not separate from but a part of nature. We love the forests, mountains, streams, oceans, and prairies. In fact, almost everyone feels, has felt, or will feel those emotions and can identify such experiences in their lives. But however eloquent we are in expressing that joy, reverence, awe—we can call it love for the nonhuman world—the ecosphere still couldn't care less about us. We find that a comforting thought. The ecosphere can't be bothered to worry about us, and for this we should be grateful, because it takes a lot of pressure off us. We don't have to pine for a love we will never get from forces greater than us.

Even though the ecosphere does not love us, it gives us everything we need to continue living, just as it gives other organisms what they need to continue living. We aren't on top. If the ecosphere favors any creatures, it would appear to be partial to bacteria, which were here before us, are here all around us and inside our own bodies, and will be here long after we are gone. As the paleontologist and evolutionary biologist Stephen Jay Gould used to say, it has never really been the "Age of Man" (using the old sexist formulation) or the age of any other multicellular organism; it is, and always has been, the Age of Bacteria.[5]

The ecosphere has given us the gift of life with no strings attached, no expectations of us. The only "rules" are the laws of physics and chemistry, which every other species has to live within as well. This is ecospheric grace: without being special or even noticed by the ecosphere, we have all that we need. The least we can do is try not to mess it up so badly. We should pay attention to those laws of physics and chemistry so as not to undermine our own species' capacity to thrive, which is simple self-preservation. And if we aren't special, we should take care not to undermine other species' capacity to thrive. It turns out that is self-preservation as well, because when we treat other species with respect we dramatically increase our ability to continue to thrive ourselves.

Why People Create God(s)

Think about this too long, of course, and our pitch to religious folks gets shaky, for an obvious reason. Most people want to be loved, want a sense of being a special kind of creature, which is likely a key motivation that leads humans to create the gods of the postagricultural world.

In Christianity, the dominant tradition talks about God the Father. Liberal Christians who reject the patriarchal construction still often conceptualize God as a parent. Our parents provide what we need to live—or, at least, should provide, though some parents either lack the resources or choose not to do so. But we expect more from our parents than just the basics needed to survive. No one would describe people as good parents simply because they made sure their kids were fed. We expect love; we need love; we cannot thrive without love. That love comes from others in the community, but the love of parents is an anchor.

If God is understood as a parent, it's inevitable that we will yearn for God's love. So if our pitch for ecospheric grace is going to work, we have to stop thinking of God as mom and dad. "Mother Earth" is an improvement over "God the Father," replacing a potentially angry parent with a nurturing one. But the problem is the yearning for parental love on a global scale. We have to recognize that while the ecosphere gives us what we need to live, it does so without guarantees. The ecosphere is not harsh—that would be projecting a human quality onto it, which makes no more sense than thinking the ecosphere loves us. But the ecosphere is not forgiving of a species' mistakes like parents tend to be.

We aren't Freudians, but Freud was onto something in his 1927 book, *The Future of an Illusion*, when he suggested "that religious ideas have arisen from the same need as have all other achievements of civilization: from the necessity of defending oneself against the crushingly superior force of nature."[6] We wouldn't make such a sweeping statement, but it's worth considering how much of the Christian tradition is born of that fear. We tend to be afraid of our vulnerability, and vulnerable people seek protection, whether from gods or gadgets. Our concern is that this fear too often leads to a denial of our vulnerability, and we take refuge in the protection we think we get from God or technological wizardry.

The ecosphere provides everything we need but doesn't love us or guarantee protection. Many people can't abide that. Enter God, a human projection of the life-giving capacity of the ecosphere but with an added feature: we're special, and God loves us even while judging our failures. We're not offering this as a full-fledged theory of God, nor are we weighing in on the question of whether humans have a religious instinct that is a result of evolution by natural selection. We are simply calling for humility. Religion both promotes humility and lays the groundwork for hubris. We're on the side of humility.

Human arrogance, what may be a kind of species predisposition to self-indulgence, is dangerous, and we want to fight it on every front. A predisposition doesn't condemn us to act out those instincts. We understand the desire for grace, for a sense of acceptance despite our flaws. But we must not let our desire for grace turn into a license for arrogance.

Ecospheric grace recognizes that we receive the gift of life from the ecosphere, for which we can be grateful, but it does not convey to us favored status. The recognition that our love for the ecosphere is not reciprocated is a reminder of our insignificance in the grander scheme—insignificant not only as individuals, but as a species. The human family is too big for any one of us to really believe we are special. The ecosphere is too big for us to really believe our species is special. There is just too much going on in this sublimely complex world to sustain that narcissistic idea.

This doesn't mean that our lives don't matter or that we can't find meaning in our lives as part of the larger ecosphere. It just means that if we need a theology—and maybe that is a human need, a yearning for meaning beyond our mere physical existence—we should create one that takes humans out of the center of the story. It's likely that throughout most of our species' time on this planet, we humans told stories that assigned meaning to our existence without that narcissism, at least until the Fall. We focus on Christianity, appropriately our subject not only because it is the tradition in which we are rooted but also because it is the religion of Europe, which for the past five centuries has been responsible for a lot of destruction. But the other monotheistic religions have much to answer for when it comes to feeding human narcissism, as do other faith traditions, especially in their more fundamentalist form.

As far as we can tell, we live in a universe that has no inherent meaning or direction, which means we are responsible, individually and collectively, for producing meaning that fulfills us and direction that guides us. This is the burden that comes with our cognitive capacity—our ability not only to ponder our own existence but also to explore the forces that make our existence possible. Yet for all that cognitive capacity, we can never be sure that we know the best way to create meaning or the best way to determine our future direction.

This produces anxiety in a fair number of people—not just today, but at least since the advent of agriculture and civilization—leading to assertions of meaning and direction that come from somewhere else, usually a god or gods but sometimes from a nontheistic spiritual source. We believe that all of these assertions are generated by people, not a product of some force, entity, or being external to people. This doesn't make any particular claim about meaning and direction inherently destructive or wrong, whether we attribute that meaning to an external force or claim it as our own. In this realm, we are philosophical pragmatists, asking whether a meaning system works. Does it help people form and maintain just and sustainable communities in which all people and nonhuman entities flourish?

Whatever one's answer to the question of the meaning of human existence, all of us should be careful about judging others before we have engaged in serious self-reflection. Scripture comes in handy to make that point as well. Jesus asks, "Why do you look at the speck of sawdust in your brother's eye and pay no attention to the plank in your own eye?" (Matthew 7:3 NIV).

In the final chapter we'll do our best to find some meaning, without pointing fingers at anyone except each other.

CONCLUSIONS

The Sum of All Hopes and Fears

When approaching a book that one wants to read but doesn't necessarily want to spend too much time on, it's tempting to turn to the last chapter and look for a summing up. This is a temptation we have felt and—more often than we would want to admit—succumbed to. The temptation seems to increase as we get older, exacerbated by living in an era of short attention spans. So if readers skipped ahead to our conclusions without reading the rest, we understand. What follows is a review of selected highlights from this book and reflections on the implications of our arguments.

The executive summary comes from the Tibetan Buddhist teacher Chögyam Trungpa: "Hope and fear cannot alter the seasons."

Scrambling for Carbon

We *Homo sapiens* have a species propensity for cooperation and division of labor, along with a symbol-generating capacity unlike any other creature. Our big brain has allowed us to create an expansive human-built world that rests with increasingly oppressive weight on top of the larger living world of which we are only a part. The weight is the result of what most people call civilization.

The material abundance generated by civilizations is the result of expansive and effective methods of capturing and storing energy—first in the soils that produced the annual grains domesticated in agriculture beginning ten thousand years ago, then in the forests used to smelt the ore of the Bronze and Iron Ages, and more recently in the coal, oil, and natural gas that have powered the Industrial and Digital Ages. That annual grain agriculture and its energy surpluses gave rise to class societies based on hierarchies, in which craftspeople, scribes, soldiers, and kings did not have to produce their own food. This led to new ideas about ownership (not just owning objects but also people, starting with men's claim to own women and children and the creation of patriarchy),[1] social systems of hierarchical control, and greater potential for expansion and conflict. The systems that institutionalized that control have varied over time and place: slave economies, feudal economies, capitalist economies.

The tapping of fossil fuels and technological advances intensified the assault on ecosystems, as a rapacious industrial capitalist system—based on the goal of continual growth and accumulation of capital without end, producing commodities of all types in huge quantities without regard for social or ecological effects—eventually dominated the planet. Food surpluses made possible by synthetic fertilizers, pesticides, and herbicides, along with dramatic advances in public health and sanitation, allowed a dramatic expansion of the human population, especially in the twentieth century.

That material abundance has never been equally or equitably distributed around the world. The widespread hierarchies that arose from elite control of energy surpluses, starting with early grain agriculture, made some people rich and kept many people poor. Movements for social justice have aimed to overcome those hierarchies and create more egalitarian relationships, which were the norm in our gathering-and-hunting past. Over millennia, these social movements that challenge the cruelty of inequality have shown us the best that human beings have to offer, our capacity for empathy and engagement. But once those hierarchies were let loose in the world, getting "civilized"—the creation of that material abundance—has shown us the worst that human beings have to offer. The idea of hierarchy has turned out to be quite tenacious, so the struggle continues.

Movements for social justice have often softened the barbarism that, ironically, has marked so much of the distinctly uncivilized imposition of civilization, and that softening is a very good thing. But the tenacity of hierarchy has been vexing. When the rock group The Who in 1971 sang,

"Meet the new boss, same as the old boss,"[2] it was not so much cynical as realistic. Countless attempts through the ages to usher in a new era of equality have ended up settling into a familiar pattern of hierarchy. This doesn't mean things never get better or that things can't get better. It's a simple recognition of reality: Hierarchy, once established, is not only difficult to dislodge but also difficult to then transcend.

Today, the ecological destruction unleashed by the invention of agriculture makes progress on social justice even more difficult. That destruction is now advanced enough that, along with our attempts under current conditions to repair the damage done to both ecosystems and human communities, we have to start thinking about how to live in a world without civilization's material abundance. In addition to struggling to distribute the wealth of the world more equitably, we have to think about that social justice project alongside the non-negotiable goal of "fewer and less": fewer people consuming less energy and material resources. The high-tech party is over. The dense energy party is over. Not everyone got invited to the party, and now we're stuck cleaning up the mess at the same time that we try to find a way to share more equitably what resources we can use responsibly.

"The arc of the moral universe is long but it bends toward justice," said Martin Luther King Jr.[3] That's a good reminder of the need for perseverance. We should also recognize that the arc of the physical universe is even longer, and it imposes limits independent of human ideas about justice. Both those observations are important, and anyone seeking social justice should not neglect the latter. If we don't continue to bend toward justice, we will have lost an opportunity to deepen our own humanity. If we don't start living within limits, we will be lucky to hold onto any sense of our humanity under the stress of collapse.

The down-powering that is necessary presents new challenges that we could potentially meet with planning that strives to be democratic and rational. But the deep denial of biophysical limits by most people in most cultures today makes such planning difficult. In some cases, the impediment is the depth of people's cult-like devotion to various systems that promise miracles. Some people believe that it is God who can be trusted to work whatever miracles are necessary for the faithful. In economics, some believe in the miracle working of capitalist markets, while others argue against markets and have faith that socialist systems and deeper democracy will somehow create sustainability. In all these social groups and ideologies, people also tend to assume that human cleverness will produce a steady flow of technological miracles that solve problems, even when that fundamentalist faith in technology created so many of the problems in the first place.

The good news is that empires not only rise but also fall, and doing what we can to bring down empires and eliminate hierarchies anywhere we find them is part of any responsible politics. A commitment to movements that aim to create a more just and equitable distribution of resources is among the most basic criteria for being a decent person. But the end of an empire doesn't guarantee the beginning of a sustainable culture. Equity and equality in economic and political realms will be a thin reed on which to lean in an era of change that is dramatic enough to warrant the term "collapse."

We must continue to struggle to overcome our capacity for cruelty, but we stand a better chance of succeeding if we acknowledge our enduring human-carbon nature. We humans, like all organisms, are engaged in the scramble for energy-rich carbon that defines life on Earth. Moral arguments and political slogans cannot change that. How we engage in the quest for carbon is shaped by ideologies—that is, by human ideas of what kind of creatures we are and how societies should work. But the carbon quest itself is rooted in biology, not ideology. Shaping an ideology that increases our chances of a decent human future requires attention to the biology.

We use "ideology" here not as a pejorative but simply to mean the set of social, political, and moral values, attitudes, outlooks, and beliefs that shape a social group's interpretation of the world. If ideology is a synonym for the framework within which people make sense of the world, then everyone has an ideology or ideologies.[4] The goal is to be self-reflective about the ideologies we articulate and adopt, in the hope that we can avoid being ideologues, people who mistake their ideology for a set of truths that are beyond challenge. To steer clear of that trap, we all should constantly ask ourselves two questions: How accurately does our framework capture the real world outside our heads, and how open are we to changing our framework in the face of new evidence?

We have tried to lead responsible intellectual lives and put those guidelines into practice. It leads us to this assessment of the current trajectory of the human species: the train is about to derail; the wheels are coming off the car; we're heading for the cliff; the waterfall is just ahead. Pick your favorite metaphor that recognizes that the future of continued endless expansion we have long imagined is over and a new future defined by contraction is coming. The unintended consequences of civilization now leave us a choice: use the big brain that makes us so clever to face honestly our problems or continue denying, minimizing, and ignoring. The former path is uncertain; the latter is guaranteed to end ugly. Species throughout Earth's history have come and gone, but we are the only one, as far as we know, that has to reflect on the possibility that the end is near. As the paleoanthropologist Ian Tattersall puts it:

> Other creatures live in the world more or less as Nature presents it to them; and they react to it more or less directly, albeit sometimes with remarkable sophistication. In contrast, we human beings live to a significant degree in the worlds that our brains remake—though brute reality too often intrudes.[5]

Brute reality is intruding all over the place these days, leaving many human ideologies stalled on the edge of the intellectual highway, which should force a reconsideration of what is possible. To borrow religious terms, we argue that people should remain rooted in a prophetic tradition that demands that we challenge the royal hierarchies that produce such suffering within the human family. But we also should be willing to speak apocalyptically, not to preach the end of the world, but to acknowledge that there is no decent human future possible within existing economic and political systems, that we are at the end of the age of those systems. The apocalyptic tradition demands that we face all the brute realities, without flinching and without false hope. The apocalyptic tradition also reminds us that life is always at the same time about death, then rejuvenation. While all the other creatures of this world will experience death, we will not only experience death but also have to ponder it. That leads many people to believe we are in a category wholly distinct from those other creatures. We think that's a mistake.

Animals or Animals+?

The first step in rational planning for the future is reviewing the past. The uncertainty of our future will be easier to accept, and the strength to persevere will be easier to summon, if we recognize the following:

- We are animals. Like all organisms, we have genetic predispositions and species propensities. Our cognitive capacities lead to a wide range of ways in which we live out those predispositions and propensities, but that range is not infinite. In some sense, humans are capable of almost anything. But human nature—and humans have a nature, just like any other organism—determines which of those almost-anythings will work over time.
- We are thinking and feeling animals. Even with our considerable rational capacities, we humans are driven by nonrational forces that cannot be fully understood or completely controlled. Even the most careful scientist is largely an emotional creature.
- We are band and tribal animals. Whatever kind of political unit a person might live in today, virtually all of our evolutionary history is in small groups of gatherers and hunters. We evolved not in cities and nation-states but in bands of typically no more than one hundred people, sometimes affiliated with other bands in a tribal structure.
- We are band and tribal animals living in a global world. The institutions of social control with formal rules that define human societies today are a relatively recent outgrowth of the invention of agriculture. We did not evolve in systems like the ones we take to be normal today, which is important to consider as we try to understand why those systems so routinely fail us. The consequences of the past ten thousand years of living as a species out of context have left us dealing with problems of social justice and ecological sustainability on a global scale that make glib talk about solutions unhelpful. With nearly eight billion people on the planet, we cannot return to gathering and hunting to survive, nor can we continue on the existing unsustainable path.

As we are forced to grapple with the complexity of our lives, we tend to forget the first point: we are animals. We are carbon-based organisms, and we would benefit from always reminding ourselves that evolutionary biology explains not only other organisms but also us. We are subject to the same laws of physics and chemistry as all other creatures. As the writer Melanie Challenger puts it, "The world is now dominated by an animal that doesn't think it's an animal. And the future is being imagined by an animal that doesn't want to be an animal."[6] But animals we are and always will be.

Today, most people—whether religious or secular—think of human beings as not simply animals but Animals+. By "Animals+," we mean that people may recognize our animal nature but, either explicitly or implicitly, believe that we are also something else, something more, something distinctly different from any other creature. Animals+ is the idea that there is something nonbiological about us that makes us special, sets us apart from all other creatures.

For most people with religious faith—not just Christianity but almost any religion—the "+" is a soul, an essence of a person beyond the physical realm that continues after death. For people who eschew traditional religion but define themselves as spiritual, the "+" often is the belief that there is an aspect of us that exists on an undefined/undefinable spiritual plane, a sense that there is something more than we can see, touch, feel. Many secular people who reject both traditional religion and nontraditional spirituality still hold onto a sense that human beings are somehow special, perhaps the only animals that have a truly free will and are not constrained by biology in the way other creatures are.

The authors don't fit any of those categories. As we have made clear, we aren't traditional believers in any religious tradition. We're not reflexively hostile to conventional religious doctrines, but they don't make sense for us. We also aren't spiritual people and, in fact, find ourselves mostly confused when people discuss spirituality. We think it's possible to experience a sense of connectedness to the larger living world without positing a nonmaterial existence. We understand that experience not as a state of higher consciousness but rather as an everyday feeling that comes with being human and fully aware of one's surroundings. And though we fit in the secular-scientific category, we are comfortable with realizing there is a whole lot about our lives that is less about choice and more about how various forces shape not only our individual life path, but our species' fate. In other words, the two of us *choose* to explore just how deeply our lives are *determined* by larger forces, recognizing the contradiction in that sentence. Such is the fate of the human animal.

The two of us think of ourselves as animals, not Animals+.[7] We try our best to be rational in our own inquiry into the world and in our evaluation of other people's theories and evidence. But we recognize that like all humans we don't fully understand our own thought processes and are subject to nonrational forces. We're not sure it makes much sense to think of rationality and nonrationality as separate processes and prefer to think of human engagement with the world as always a complex mix. Facing "the combat of passion and of reason," we grapple with David Hume's famous conclusion: "Reason is, and ought only to be the slave of the passions, and can never pretend to any other office than to serve and obey them."[8] We wish reason could get the upper hand, but we recognize what kind of animals we are. Still, we continue to act as if our passions can be directed by the conclusions of reason, recognizing the contradiction in that position. We strive to be more clear-eyed in our use of reason, knowing how powerful the passions are. As the radical activist Abe Osheroff used to say, passion is the engine of human history, but when we board that train we should remember to bring along our carry-on luggage of critical reasoning.[9]

In summary: We are not sectarian in religion, we are not spiritual in inclination, and we reject any secular version of Animals+. We recognize the limits of reason, but we prefer it to claims made through revelation. In philosophical terms, we are monists, not dualists: we think all the stuff in the world is made of the same kind of stuff and don't see a need for concepts such as the soul. We are what most philosophers would call materialists, or believers in physicalism, which is to say, the one kind of stuff there is in the world is physical. One way to make these terms tangible: When people talk about "the mind," we think of "the brain." Another: When our biological processes in the brain and body finally stop—no more blood flowing, no more synapses snapping—we are dead and done. Our body begins the process of returning to the earth and we live on only in the thoughts of those who knew us or knew of us.

We realize there is a vast literature in philosophy, neuroscience, and many other fields relevant to these assertions, and we don't pretend to be experts in those debates. Our conclusions are based on two lifetimes of engagement with the world, practical and scholarly, articulated in a way we hope other nonspecialists can appreciate. Some would suggest that these positions indicate we lack imagination, that we can't see beyond the obvious and superficial aspects of our existence. We disagree, based not just on an intellectual argument, but on our experience.

Both of us have the capacity to love other people deeply—in the romantic-intimate sense, as well as in myriad other ways that defy easy categorization—in a fashion that we don't fully understand. Love is, indeed, a mystery.

We can reflect on our emotional lives and recognize that whatever research and reason might reveal about human psychology, we barely know ourselves as individuals or as a species. Our own selves are, indeed, a mystery.

We cannot help ourselves from pondering questions that people likely have pondered for as long as there have been people. Why is there something rather than nothing? What came before the universe came into being? We have no expectation of definitive answers but great expectations of a pleasurable engagement with others in speculating.

And, like most people, when out at night under the stars, we can come up with no better description of the sensation than the commonplace: a sense of wonder.

All of this is a source of endless vexation and immeasurable joy. We wish we understood it all, and we are glad we don't.

"Why Is This Not Enough?"

Jackson was out walking on his property in Kansas one spring day and called Jensen with a simple question: "Why is this not enough?"

Jackson ticked off a list of the plants he had cataloged on the walk and described a spider web between two trees that he had been studying. Then he talked about imagining the journey that a branch he threw into the Smoky Hill River that runs behind his house would take to the

Gulf of Mexico. The Smoky Hill joins the Republican River at Junction City, which forms the Kansas River, which joins the Missouri River at Kansas City, which joins the Mississippi River just north of St. Louis, which flows into the Gulf of Mexico. Although the Mississippi is the more famous of the two rivers, the Missouri is actually longer, the longest river in North America at 2,341 miles, 21 miles longer than the Mississippi. As Jackson likes to say, we threw in those details at no extra charge.

Back to the branch. Would that branch make it all the way to the Gulf? Would it get hung up on the shore at some point? How many people would we have to deploy to watch along the banks of the rivers to get an authoritative account of the branch's journey? How broken up on rocks would it have to get before we would say it is no longer the same branch but would instead consider the broken-up parts new things, a naturalistic version of the ancient Ship of Theseus thought experiment (if, board by board, every component of the ship were eventually replaced, would it still be fundamentally the same ship?). Those are the kinds of questions that go through children's minds when they throw a branch into the river. If we are lucky, they are the kinds of questions we still ponder as adults. They are the questions that arise from an active imagination.

"Why is this not enough?," Jackson asked. Why are the sights and smells of the world, along with the questions that the world generates, not enough for us humans? Why are our everyday questions about the world, which can help us appreciate the world more, not enough? Do we really need male sky gods? Do we really need to dream about colonizing Mars? What if sky gods and Mars colonies are the sign of an atrophied imagination?

Indeed, why is this not enough?

We find the authentic underpinnings of meaning in our own engagement with the world around us. All of this—the seeing, the smelling, the pondering, the emotions that arise in us when we talk about it all—is enough.

Enough for what? For us, it's enough to create an adequate sense of the meaning that humans seem to require. If the physical world is all there is, so what? It's a pretty amazing place. It's still a world in which our attempts to understand can provide all the meaning we need.

We believe this approach to being human on this planet is enough to get us started on a response to the crisis of meaning we wrote about in our introductions. That's the easy part.

Crisis of Consumption

Much more challenging is the crisis of consumption, for at least two reasons. First, consumption in the affluent world does provide a certain kind of meaning. The pursuit of goods and services can provide at least an immediate sense of fulfillment. Even if the rewards of that pursuit are inadequate over the long run, those products and perks provide a sense of comfort, security, and superiority. If we have enough money to spend freely on ourselves, we can easily feel that our needs and wants have been met, that we're safe in the world, and that we are better than those who don't have the means to consume so lavishly. Some of us may not like that way of finding meaning, but it is meaning of a sort.

Of course, when stated like that, no thoughtful person would say, "Yes, that's all the meaning in life I need." But in practice, those rewards of affluence can quiet the internal voice that reminds us that consumption is never really enough for anyone. The meaning that comes from affluence is seductive because it's easy, because it asks nothing from us but to conform to the values of a system that has no values beyond greed and acquisition. It's fragile, but as long as the money keeps rolling in, many people can stay afloat that way.

Second, it's not easy to stop consuming. In chapter 4, we spoke of the way that wants become needs, that luxuries become necessities, that once we have access to certain goods and services it is not easy to simply give them up. Even people who would say that the deeper meaning they have in their lives does not come from consumption find it hard to give up consuming at the level to which they have become accustomed. We aren't saying it never happens or can't happen. We are saying it's not easy, as we know from experience.

These struggles of the affluent likely sound petty to people living in deep poverty or living paycheck to paycheck with no money for anything beyond real necessities. To people with so little, it must seem odd that people with a lot find it so difficult to give up some portion of "more than I need." It seems that way to us too, that it should be easy. But it isn't, and we don't have to look to the self-indulgent superrich for evidence. All we need do, as we pointed out in chapter 4, is reflect on our own lives. And though it may seem churlish to point out, many people who make their way out of poverty end up embracing the easy but empty answers of affluence.

The problem is captured succinctly, once again, by the nineteenth-century Austrian writer Marie von Ebner-Eschenbach in her book of aphorisms: "To be satisfied with little is hard. To be satisfied with a lot is impossible."[10] We both have known people who were better than we are at resisting that, and we think of them often. For Jackson, the one person who took the aphorism most seriously was Leland Lorenzen.[11] For Jensen, it was Jim Koplin.[12] Both of those friends are now dead, but part of what keeps them in the front of our minds constantly is their commitment to a frugality that is unusual in the affluent sectors of the United States. In both cases, that frugality wasn't a sacrifice. Leland and Jim described the way they lived as the easiest way to move through the world, which made them two of the freest people we have met. Neither was motivated by self-righteousness. As Leland told Jackson, "I do this because it's easy." Giving up most material possessions and living in a small shack wasn't a hardship for Leland but a relief.

We admire them, but we aren't holding them up as a solution. The exception proves the rule once again: they were quirky people, unconcerned with how others judged their choices. We can't assume everyone will follow the few people who choose frugality. After all, neither of us has so far lived up to their standards. We can't ignore the fact that the vast majority of people who get their hands on a little will seek the satisfaction of getting a lot, then more than a lot. We can't ignore the fact that a capitalist consumer economy spends billions of dollars every year on advertising and marketing to reinforce that tendency. We can't build a world on the example of Leland and Jim, even though in the future more and more people will have no choice but to embrace such frugality.

An important caveat here: We are talking about a frugality that can enrich everyone's lives, not austerity policies that are imposed by people with wealth and power on people already struggling. When a country runs into financial problems these days, those wealthy and powerful folks love to preach what the economists call austerity: cuts in what is often already an inadequate social safety net, reductions in government spending that help the poor, and a call for ordinary people to tighten their belts. We recognize that we will all face a belt-tightening future whether we like it or not, but that is not an endorsement of the cruel austerity policies in contemporary policy making.

So, at risk of excessive repetition, we repeatedly repeat: the existing distribution of wealth and power is morally indefensible. Struggles to achieve justice today matter. But there will be a time when even the financial instruments hoarded by the wealthy, and the political power such wealth can buy, will offer little privilege or protection.

Here's a way to sum up the quandary we are in:

- If eight billion people lived at the level of an average American, the modern economy would exhaust all of the world's ecosystems quickly. It would be "game over" today.
- If eight billion people reduced their consumption to the level of an average European, the same result would take slightly longer. Game over but a bit later in the week.
- If eight billion people lived at the level of an average citizen of the developing world, the same result would take longer yet. Game over later this year, perhaps.

There are cultures and systems that have intensified this crisis, capitalism and European imperialism being the most recent. Taking the long view, we see that today's unjust systems are rooted in patriarchy[13] and the class/status hierarchies that emerged as humans settled into life based on the domestication of plants and animals and the surpluses that agriculture produced. But at the heart of the problem is human-carbon nature. The carbon-seeking behavior of all creatures became potentially ecocidal when the cognitive capacity and cooperative nature of *Homo sapiens* dramatically increased our capacity to capture energy.

Our conclusion:

- Capitalism's demand for endless growth is incompatible with a livable human future, but
- there is no economic or political system, no matter how democratic or just, that can deliver even a relatively modest version of the lifestyles generated in fossil-fueled capitalism, which is a way of saying that
- first world levels of consumption are unsustainable no matter how the world is organized, and reorganizing the world along criteria of ecological sustainability is going to be difficult because
- the modern industrial world and a global consumer culture have generated expectations that can't be met, and coming to terms with that will be difficult because
- the Industrial and Digital Revolutions have not only changed our relationships to each other and created unrealistic material expectations; they have fundamentally changed the way most people interact with the nonhuman world.

All this brings us back to an observation that Jackson has been making for decades: we are a species out of context.

Jensen has long appreciated Jackson's ability to capture an important idea in a single phrase or simple sentence, and he keeps telling Jackson that "species out of context" is his most important. Of course, any good anthropologist knows that we no longer live in the kinds of societies in which we evolved as a species. In fact, any reasonably attentive high school student knows that. But we rarely ponder the implications of our species-out-of-contextness.

Here's one way to make this point. Our genus, *Homo*, has been around for a little more than 2.5 million years. Our species, *Homo sapiens*, has existed for about 200,000 to 300,000 years. Our journey as *Homo hierarchicus*—the humans who created widespread systems of domination and subordination starting with agriculture and patriarchy—is but a blink of an eye in evolutionary terms.[14] Even more recent is the emergence of *Homo colossus*,[15] the humans living in the industrial world who use fossil fuels to consume at levels previously unimaginable. Now we live in the age of *Homo technologicus*,[16] the humans who believe that our own cleverness and knack for invention can keep our ecosystem-destroying activity going indefinitely.

Here's one way of summarizing contemporary failures. *Homo technologicus* is trying to solve the dramatic problems created by *Homo colossus* without acknowledging that the seeds of destruction were planted by *Homo hierarchicus.* We are living with illusions that are not only the product of today's irrational technological exuberance and the past two centuries of fossil energy gluttony but ten thousand years of drawing down the ecological capital of Earth beyond replacement levels.

The intensification of that drawdown over the past century has left us no alternative but to plan for a major down-powering of human societies. Whether we like it or not—whether we can imagine it or not—the future will be fewer people consuming less.

Today's crises are best understood in that context. Corrupt and craven leaders are a problem but not the main problem. Unjust and unsustainable systems are a problem but not the core problem. To face the multiple cascading crises that define this historical moment requires confronting our human-carbon nature and the forces unleashed when we became a species out of context.

Making judgments about the failures of individuals and systems, and acting on those judgments, is not only understandable but a political obligation. Different people will start with different assumptions and offer different evaluations. There is, after all, a lot of variation in the human species. Because people will come to different judgments, there will be conflict, and there are better and worse ways of resolving that conflict. We prefer nonviolent approaches based on rational discourse, recognizing that we all are driven by passions that cannot, and should not, be ignored. We seek approaches grounded in widely accepted moral principles rather than those that are explosively emotional. But we all are too deeply embedded in large-scale hierarchies to imagine there is an easy path that everyone agrees will get us to the small-scale and egalitarian communities that are our evolutionary context.

Is There Hope?

We have throughout this book tried to confront honestly the problems we humans face. Both of us have been writing and talking in public in this fashion for decades, and the most common questions in response, typically the last question at the end of a lecture or conversation, concerns hope. The two most common variations are, "Is there any reason to have hope?" and "How do you sustain hope in your life?"

Many other writers have engaged these questions without sugarcoating the facts or floating off into platitudes. For example, Joanna Macy and Chris Johnstone offer the concept "active hope" as a practice.[17] The focus is on actions that can help advance a collective transition, what Macy calls a "Great Turning" toward justice and sustainability.[18] The primatologist Jane Goodall and the writer Douglas Abrams say that hope "is what enables us to keep going in the face of adversity," allowing us to articulate the future we want to see, but that "we must be prepared to work hard to make it so."[19] Hope is not something that one person can give to another. The value of the idea is in a call for collective action.

A powerful vision of this kind of strength appears in Octavia Butler's novels *Parable of the Sower* and *Parable of the Talents*, in which the protagonist, Lauren Olamina, creates a new religion of sorts to help people cope with intensified suffering and injustice during a period of collapse. Her "Earthseed" philosophy recognizes change as inevitable and posits God as Change.[20] We agree that embracing change is psychologically healthy.

Back to those questions. On the first formulation, the obvious clarifying question is, "Hope for what outcome?" As we have made clear, we have no hope that eight billion people can live on Earth in anything like the current social, economic, and political systems. If that's the goal, then we counsel giving up. Better to articulate new goals that are consistent with what we know about ecology and basic biology, operating within the biophysical limits set by the laws of physics and chemistry. Let's say that the new goal is getting to a world of two billion people who consume far less energy and resources. Do we have hope that our species can get there, with as little human suffering and as little ecological destruction as possible?

Yes, we do believe it is possible to limit the suffering and destruction, but that is a feasible goal only if in addition to a critique of the forces that have created so much social injustice—patriarchy, white supremacy, capitalism, imperialism—we also recognize that we all share the same human-carbon nature. The possibility of a human future depends just as much on coming to terms with the temptations of dense energy that are so hard to resist because of our nature. A collectively imposed cap on carbon seems to be not only the potentially most effective approach but likely the only one that could work.

On the second formulation of the question, we decided to write separately, to describe our different experiences. As we said in our introductions, we are similar in many ways and different in many ways. This is one place where the differences matter, a place where we recognize that our experiences are crucial to how we move through the world.

Hope is Irrelevant

Robert Jensen

How do I sustain hope? I don't, because I can't sustain what I don't have and never had. Hope has never been terribly relevant in my life, never been a big part of my motivation to act in the world.

In my writing and political projects—from my first work in the radical feminist movement to challenge men's sexual exploitation of women in pornography[21] through organizing to challenge US imperialism and militarism[22] to current work on the ecological crises—I typically have assumed that whatever positive change was likely to result from my efforts would be small, and even small changes would not be guaranteed. At the same time, I have tried my best to engage in efforts that could achieve as much as possible, reach as many people as possible, and try to be as strategic as possible to achieve realistic short-term goals while still focusing on the need for radical change over the long term.

As a professor with a steady income, perhaps it was easy for me to take that view, to keep working without much hope of significant success. At the end of the day, I always had a roof over my head, I ate regular meals, and I could go to the doctor if I got sick. In material terms, my adult life has been comfortable by any reasonable standard. That's important to acknowledge, but lots of people with similar privilege do not find it so easy to work without much hope. Hope seems to be necessary for people across the spectrum. Why has a kind of joyful hopelessness been second nature for me?

While introspection is not a perfect method for answering such questions, here's my best guess. My early experience in the world was defined by trauma, on multiple levels from multiple sources, fairly relentless and with no safe harbor. It's not necessary to recount the details, but that broad outline is relevant to why I am so tone-deaf to talk about hope. Long before I was capable of understanding the forces that produce such trauma—not only for me, but for countless others—I had to live with it, without support and with no expectation of better days ahead. I was lucky eventually to have opportunities for higher education and satisfying professional work, but by that time I had found a way to live that did not require hope.

I have gravitated toward projects for social justice and ecological sustainability because they have provided some meaning in my life, and by the time I made those choices I had concluded that the only meaning in our lives is created through our own thoughts, words, and deeds. I don't recall ever searching for the divine or seeking epiphanies to provide meaning. Instead I developed a rather banal workaday attitude: get up in the morning, day after day, try to find something worth doing, and then do it as well as possible, realizing that failure will be routine but that small successes—sometimes really small, maybe even too small to see in the moment—make it possible to continue.

The systems that govern the world have demanded that I give a fair amount of my time and energy to a boss. Like most of us, I have had to meet the demands of various employers so that I can pay my bills and live a kind of normal life. But I have carved out as much space as possible for activities that challenge me personally and intellectually. I have sought the company of others who also seek those challenges. I have tried to create opportunities to help remedy problems in whatever small way possible.

I have done this not out of hope for dramatic change in the world but because it has been for me the best way to live a decent life. Positive change happens, of course, and should be celebrated, even when it's the clichéd pattern of two steps forward, one step back. Even when it's two steps forward and three steps back, we can take a step to the side to try another route. Creative responses to rejection and failure are always possible.

People have told me that this approach is a kind of hope in itself, that I have found hope in the way I abandoned hope. At that point, the words we choose don't matter much. What does matter is getting out of bed in the morning and finding work that is worth doing. I believe in this path not just because it has sustained me, but because I have seen it sustain others, and sharing this perspective with others has made it possible for me to plod forward.

I'll end with a story about one of those people, whom I've already described as being especially important to me, Jim Koplin. After he died, I wrote a book about him, but here I'll tell one story that's not in that book. Jim's early experience was also defined by trauma, and his struggles to live with those harsh realities resonated with me. Growing up as an only child on a Depression-era Minnesota farm, Jim spent a lot of time alone. As an intellectually minded kid, he spent a lot of time reading and in self-reflection. He told me that at one point as a child, he realized that every person on Earth had basically the same cognitive and emotional capacities as he did—that we were pretty much all the same kind of creature. That meant that every person had the same capacity as he had to feel pain and to suffer. The suffering he and his mother endured at the hands of an abusive father was considerable, and he knew from reading that others around the world suffered as much, sometimes much more. The awareness of the scope of pain in the world overwhelmed him, so he took the family rifle out to the woods with the intention of killing himself. He sat alone in the woods for some time before deciding to live and return home. But from that point forward, he said, he knew that he had to find ways to acknowledge the pain of the world but also insulate himself from a constant awareness of it, or he would not survive.

He not only survived, but thrived. Until his death at age seventy-nine, Jim was committed to radical political activity and loving community connection. I was fortunate to know him for his last twenty-four years, and I now am part of a large circle of friends of Koplin, people whose lives were changed by his quiet commitment to decency, by watching him honor the dignity of others. He was the first person who talked to me about the grief that was inevitable if we told the truth about the world, and he remains my model for being honest with myself and others.

Did Jim Koplin have hope? I don't recall the word ever coming up in our many conversations about these subjects. Jim simply got out of bed in the morning, tended his garden, volunteered with community groups of all kinds, showed up at rallies and protests, laughed with his neighbors, struggled with his own unresolved demons, and went to bed early so that he could get up early to do it all over again.

If that's hope, so be it. Whatever we call this approach to life, it's more than enough to get me up in the morning.

Hope in a Ponzi Scheme?

Wes Jackson

Over the years, people have often put the question to me, "Are you an optimist or a pessimist?" In years past, I usually responded with something like, "I'm not an optimist, but I am hopeful." I would suggest that the reasons for hope have to do with the good work of many people in numerous organizations who have dedicated so much of their lives to justice and sustainability.

I understand the intent of the question, and that answer is perfectly adequate, I suppose. But it never felt like enough, and I never came up with a better answer. In recent years, I have done my best to avoid using the word *hope*.

The reason has to do in part with a growing recognition that we are all caught in one big Ponzi scheme that started with agriculture. A Ponzi scheme is a con in which investors are paid not with the profits from successful investments or products but with money from a steady stream of new investors. Such scams didn't start with Charles Ponzi, but his name stuck because of the large amount of money he took in: $20 million in the early 1920s, which is at least ten times that much in inflation-adjusted dollars today. Bernie Madoff set the record for Ponzi scams by taking in tens of billions of investors' dollars and making off with an estimated $20 billion for himself.

Madoff died in prison while we were working on this book, and it got me thinking about human carbon seeking as a kind of Ponzi scheme, starting ten thousand years ago with agriculture. Although it began without fraudulent intent, the human pursuit of carbon is the biggest Ponzi scheme of them all. When the first farmers domesticated wheat in what we now call the Middle East, they had no way of knowing that they were creating the surpluses that would lead to empires, that those empires would exhaust soils around the world as they rose and fell, or that eventually nearly eight billion people would be dependent on keeping that carbon-extraction scheme going.

But here we are now, stuck with our place near the end of the long line of Ponzi investors. People have been drawing down the ecological capital of the ecosphere ever since agriculture, taking from Earth in ways that reduce options for future generations. As we have emphasized throughout this book, the surpluses haven't been equitably distributed and not everyone has been enriched along the way. But now the human species, collectively, has to face the consequences of the unintentional scam but with no Charles Ponzi or Bernie Madoff to be the target of our anger.

Back to hope. If this intensified carbon seeking kicked off by agriculture is a kind of Ponzi scheme, paying off one generation by drawing down the assets of the next, what kind of hope makes sense? Ponzi schemes only end one way, with some people paying for the illusion of other people's embrace of endless growth and high returns. Where's the hope if that's what we're up against?

Well, out of the same region where people first farmed, there eventually came new religions, including Christianity, eventually in many versions, including the faith of my mother. From the apostle Paul, the guy who wrote letters trying to get new Christians to fly right, we learn about faith: "Now faith is the substance of things hoped for, the evidence of things not seen." That's from his letter to the Hebrews 11:1, the King James Version. For most of that chapter, Paul explains how people in the stories of the Hebrew Bible who had faith in God were rewarded. But in the last few verses, he acknowledges that some people with faith also ended up wandering, persecuted, stoned to death. Paul isn't promising what can't be promised. For him, faith is no guarantee of success in this life, only of an eventual resting place with God.

I grew up in a community in which many, maybe most, people believed that a final reward was waiting for the faithful. I'm no longer one of those believers, but I certainly agree that there are no guarantees on this Earth. That's truer than ever, given the short time frame in which we must act. Scientists say we need to cut the use of carbon in half this decade if we are to have a chance at staving off the worst consequences of climate change. A tall order, indeed. With that on the table, what kind of faith makes sense?

I don't have a grand answer to that. It seems to me that what we can do is get busy and do our best to deserve a soft landing.

THE TWO OF US, JACKSON AND JENSEN, are a lot alike, and we take great pleasure in that. We also are dramatically different, and both of us have learned much from that. From different beginnings, we have ended up in a similar place philosophically. We both count ourselves lucky for the late-in-life unexpected pleasure of it all. Trying to explain such good fortune is difficult. We recognize our limits, so we will give the last word on hope in this book to Jackson's longtime friend Wendell Berry, who weighs in on the subject in one of his Sabbath poems.

> It is hard to have hope. It is harder as you grow old,
> for hope must not depend on feeling good
> and there is the dream of loneliness at absolute midnight.
> You also have withdrawn belief in the present reality
> of the future, which surely will surprise us,
> and hope is harder when it cannot come by prediction
> any more than by wishing. But stop dithering.
> The young ask the old to hope. What will you tell them?
> Tell them at least what you say to yourself.[23]

If we have any advice for others, maybe those lines capture it better than we can through analysis. When struggling to know what to say to others about our own fears, we can tell them at least what we say to ourselves.

If there is a path forward, it requires us to face our worst fears of the future and then speak of them fearlessly. In this book, we have tried to make good on that challenge. This book is what we say to each other, and it is what we say to ourselves.

Our final summing up: Since the invention of agriculture, since "the Fall," hubris has led humans to believe that we can play God, to assume that human cleverness is adequate to run the world according to our dictates. That project has failed, and our most important asset going forward is humility. Our arrogance in grabbing from the Tree of the Knowledge of Good and Evil cannot be undone, but we can do our best to make our peace with the Tree of Life.

That advice applies not only to those most directly responsible for the mess we're in, but to everyone. Because we remain fully committed to the idea of one human species, rejecting racism and ethnocentrism, we remind ourselves that *the moral high ground is a dangerous place to stand*, even when standing there might be warranted.

We humans are not godlike in our ability to know good and evil, but we can do our best to understand the signals that the ecosphere is sending and act intelligently, for our own self-respect and for the sake of the planet's creatures, human and other. Our species propensity for cooperation, combined with our cognitive abilities and symbolic capacities, has gotten us into trouble. But those same attributes are also available to help us atone. We are stuck using the same big brain that brought us to this place in history to try to prevent more suffering, lessen the destruction, and create a soft landing after existing social, political, and economic systems are gone, either because we worked to transcend them or because of collapse. We are stuck using the assets that got us in trouble to try to get out.

We're tempted to end on a cheeky note, by suggesting that if we are wrong and there is a god, it is the God of Irony. In the world of that god, there is no Promised Land to find, no Golden Age to return to, no Rapture to await. There is simply the hard work, in front of us every day, of reconciling ourselves to the limits that the larger living world imposes. There is the struggle to do that work, with gratitude for the grace that world offers.

And there is love. In 1 Corinthians 13:13, the apostle Paul tells us that of faith, hope, and love, the greatest is love. Along with the love among humans, which can be reciprocated, we have highlighted people's love for the larger living world around us. But the ecosphere does not return that affection and doesn't care about us. We have to remember that always. When facing difficult truths, it's tempting to want to slide out of trouble with an invocation of love, and there's nothing wrong with that so long as the invocation doesn't become a mode of evasion. We should not use love to avoid harsh realities but rather to help us face them. Harshness often flows from anger and resentment, but harshness also can come out of love. Dostoyevsky said it better than we can:

> If you do not attain happiness, always remember that you are on the right road, and try not to leave it. Above all, avoid falsehood, every kind of falsehood, especially falseness to yourself. Watch over your own deceitfulness and look into it every hour, every minute. Avoid being scornful, both to others and to yourself. What seems to you bad within you will grow purer from the very fact of your observing it in yourself. Avoid fear, too, though fear is only the consequence of every sort of falsehood. Never be frightened at your own faint-heartedness in attaining love. Don't be frightened overmuch even at your evil actions. I am sorry I can say nothing more consoling to you, for *love in action is a harsh and dreadful thing compared with love in dreams.*[24]

Love does not make the world go 'round—that's physics—and the world will go 'round without us. But love is essential to how we humans go 'round in this world for as long as we have left. We end on what we believe is the most upbeat note possible, offering a strong endorsement of love in action, recognizing that at our moment in history that love is bound to be harsh and dreadful. But it is still love.

NOTES

Introductions

1. John Steinbeck, *Log from the Sea of Cortez* (New York: Penguin, 1995), 213.

2. "Ecosphere" is the term we prefer for what some call the biosphere, defined as the space for life on Earth. The global ecological system is created by organisms' interactions with the lithosphere (Earth's crust and upper mantle), hydrosphere (water on the surface, as well as above and below that surface), cryosphere (water in solid form), and atmosphere (the layer of gases retained by Earth's gravity).

3. This phrase, which we will use often, comes from the late Jim Koplin, a longtime friend of Jensen. See Robert Jensen, *Plain Radical: Living, Loving, and Learning to Leave the Planet Gracefully* (Berkeley, CA: Counterpoint/Soft Skull, 2015).

4. Carl Sigman and Bob Russell, "Crazy He Calls Me" (1949). The most well-known recording of the song is by Billie Holiday.

5. T. C. Chamberlin, "The Method of Multiple Working Hypotheses," *Science* 148, no. 3671 (1965): 754–59.

6. T. S. Eliot, "East Coker," in *The Four Quartets* (1940). Accessed at www.davidgorman.com/4quartets/2-coker.htm.

7. "World Scientists' Warning to Humanity," available at www.ucsusa.org/resources/1992-world-scientists-warning-humanity. Henry Kendall, a Nobel Prize–winning physicist and former chair of the Union of Concerned Scientists' board of directors, is the primary author.

8. William J. Ripple et al., "World Scientists' Warning to Humanity: A Second Notice," *BioScience* 67, no. 12 (December 2017): 1026–28, https://academic.oup.com/bioscience/article/67/12/1026/4605229.

9. Commission for the Human Future, "Surviving and Thriving in the 21st Century" (2020), https://humanfuture.net/sites/default/files/CHF_Roundtable_Report_March_2020.pdf.

Chapter One. Who Is "We"?

1. F. Scott Fitzgerald, "The Crack-Up," *Esquire*, February 1936, www.esquire.com/lifestyle/a4310/the-crack-up/.

2. Richard Levins and Richard Lewontin, *The Dialectical Biologist* (Cambridge, MA: Harvard University Press), 141.

3. An early version of some of these ideas appeared in Robert Jensen, "Who Is We?," *Ecological Citizen* 4, no. 1 (2020): 57–61, www.ecologicalcitizen.net/pdfs/v04n1-11.pdf.

4. Md Arif Hasan and Ralph Brougham Chapman, "The Environmental Footprint of Electric versus Fossil Cars," The Conversation, October 15, 2019, https://theconversation.com/climate-explained-the-environmental-footprint-of-electric-versus-fossil-cars-124762.

5. System Change not Climate Change: An Anti-capitalist, Ecosocialist Network, https://systemchangenotclimatechange.org/.

6. Lisi Krall, *Bitter Harvest: An Inquiry into the War between Economy and Earth* (Albany: SUNY Press, 2022); Lisi Krall, "The Economic Legacy of the Holocene," *Ecological Citizen* 2, no. 1 (2018): 67–76, www.ecologicalcitizen.net/pdfs/v02n1-11.pdf.

7. See, respectively, Giorgos Kallis, Susan Paulson, and Giacomo D'Alisa, *The Case for Degrowth* (Medford, MA: Polity Press, 2020); Jason Hickel, *Less Is More: How Degrowth Will Save the World* (London: Windmill Books, 2021); Herman E. Daly, *Steady-State Economics*, 2nd ed. (Washington, DC: Island Press, 1991); Kate Raworth, *Doughnut Economics: Seven Ways to Think like a 21st-Century Economist* (White River Junction, VT: Chelsea Green Publishing, 2017).

8. Johan Rockström and Mattias Klum, with Peter Miller, *Big World, Small Planet: Abundance within Planetary Boundaries* (New Haven, CT: Yale University Press, 2015); "Planetary Boundaries," Stockholm Resilience Centre, www.stockholmresilience.org/research/planetary-boundaries.html.

9. For an expression of this optimism that borders on self-parody, see "The Sky's the Limit: Solar and Wind Energy Potential Is 100 Times as Much as Global Energy Demand," Carbon Tracker, April 23, 2021, https://carbontracker.org/reports/the-skys-the-limit-solar-wind/.

10. Laura J. Sonter, Marie C. Dade, James E. M. Watson, and Rick K. Valenta, "Renewable Energy Production Will Exacerbate Mining Threats to Biodiversity," *Nature Communications* 11, no. 4174 (2020), www.nature.com/articles/s41467-020-17928-5; Éléonore Lèbre, Martin Stringer, Kamila Svobodova, John R. Owen, Deanna Kemp, Claire Côte, Andrea Arratia-Solar, and Rick K. Valenta, "The Social and Environmental Complexities of Extracting En-

ergy Transition Metals," *Nature Communications* 11, no. 4823 (2020), www.nature.com/articles/s41467-020-18661-9.

11. Michael Klare, "Lithium, Cobalt, and Rare Earths: The Post-Petroleum Resource Race and What to Make of It," TomDispatch, May 20, 2021, https://tomdispatch.com/lithium-cobalt-and-rare-earths/.

12. Simon P. Michaux, "The Mining of Minerals and the Limits to Growth," Geological Survey of Finland, https://tupa.gtk.fi/raportti/arkisto/16_2021.pdf.

13. Steven K. Beckner, "Admin Doesn't Jawbone the Fed, Does It?," Market News International, June 24, 1996.

14. Alan Goodman, "Race Is Real, But It's Not Genetic," *SAPIENS Anthropology Magazine*, March 13, 2020, www.sapiens.org/biology/is-race-real/.

15. David J. Linden, *Unique: The New Science of Human Individuality* (New York: Basic Books, 2020), 247; emphasis in original.

16. R. A. Foley and M. Mirazón Lahr, "The Evolution of the Diversity of Cultures," *Philosophical Transactions of the Royal Society B* 366, no. 1567 (2011): 1080–89, www.ncbi.nlm.nih.gov/pmc/articles/PMC3049104/.

17. Alfred W. Crosby, *Ecological Imperialism: The Biological Expansion of Europe, 900–1900*, 2nd ed. (Cambridge: Cambridge University Press, 2004); Jared Diamond, *Guns, Germs, and Steel: The Fates of Human Societies* (New York: Norton, 1997); Ian Morris, *Why the West Rules—for Now: The Patterns of History, and What They Reveal about the Future* (New York: Picador/Farrar, Straus and Giroux, 2010).

18. Alice J. Friedemann, *Life after Fossil Fuels: A Reality Check on Alternative Energy* (Cham: Springer Nature Switzerland, 2021).

19. Various ecologists have formulated versions of this "maximum power principle." See Howard T. Odum and Elisabeth C. Odum, *A Prosperous Way Down: Principles and Policies* (Boulder: University Press of Colorado, 2001), 69–71. Richard Heinberg suggests a modification to respond to the ecological crises, emphasizing the need to be willing to sacrifice some uses of power in the present in order to maximize power over a longer time, what he calls an optimum power principle. Richard Heinberg, *Power: Limits and Prospects for Human Survival* (Gabriola Island, BC: New Society Publishers, 2021), 292.

20. David Despain, "Early Humans Used Brain Power, Innovation and Teamwork to Dominate the Planet," *Scientific American*, February 27, 2010, www.scientificamerican.com/article/humans-brain-power-origins/.

21. Reconciling the two books is sometimes called "the Adam Smith problem." See Keith Tribe, "'Das Adam Smith Problem' and the Origins of Modern Smith Scholarship," *History of European Ideas* 34, no. 4 (December 2008): 514–25, www.tandfonline.com/doi/full/10.1016/j.histeuroideas.2008.02.001. See also Blake

Smith, "Adam Smith Warned Us about Sympathising with the Elites," *Psyche*, October 5, 2020, https://psyche.co/ideas/adam-smith-warned-us-about-sympathising-with-the-elites.

22. Adam Smith, *The Theory of Moral Sentiments*, pt. 1, "Of the Propriety of Action," ch. 1, "Of Sympathy" (1759), www.gutenberg.org/files/58559/58559-h/58559-h.htm.

23. Lydia Syson, "The Radical Aristocrat Who Put Kindness on a Scientific Footing," *Psyche*, October 26, 2020, https://psyche.co/ideas/kropotkin-the-radical-aristocrat-who-put-kindness-on-a-scientific-footing.

24. Peter Kropotkin, *Mutual Aid: A Factor of Evolution* (1902), https://theanarchistlibrary.org/library/petr-kropotkin-mutual-aid-a-factor-of-evolution.

25. Peter Kropotkin, *Ethics: Origin and Development* (1921), ch. 2, "The Gradually Evolving Bases of the New Ethics," https://theanarchistlibrary.org/library/petr-kropotkin-ethics-origin-and-development.

26. Kim Sterelny, "How Equality Slipped Away," *Aeon*, June 10, 2021, https://aeon.co/essays/for-97-of-human-history-equality-was-the-norm-what-happened.

27. Elinor Ostrom, "Beyond Markets and States: Polycentric Governance of Complex Economic Systems," Nobel Prize Lecture, December 8, 2009, 435–36, www.nobelprize.org/uploads/2018/06/ostrom_lecture.pdf.

28. Vitek, a philosopher by training, directs the New Perennials Project and New Perennials Publishing. See www.newperennials.org/.

29. Robert Jensen, *The Restless and Relentless Mind of Wes Jackson: Searching for Sustainability* (Lawrence: University Press of Kansas, 2021), 27–29.

30. Ibid., 29–31.

31. System Change not Climate Change, "Points of Unity," https://systemchangenotclimatechange.org/points-of-unity/.

32. Joseph A. Tainter, "Energy, Complexity, and Sustainability: A Historical Perspective," *Environmental Innovation and Societal Transitions* 1, no. 1 (2011): 89, www.sciencedirect.com/science/article/abs/pii/S221042241000002X.

33. Stan Cox, *The Green New Deal and Beyond: Ending the Climate Emergency While We Still Can* (San Francisco: City Lights, 2020); Stan Cox, *Any Way You Slice It: The Past, Present, and Future of Rationing* (New York: New Press, 2013).

34. Kirkpatrick Sale, *Rebels against the Future* (Reading, MA: Addison-Wesley, 1995), 261.

35. Robert Jensen, "The Danger of Inspiration: A Review of *On Fire: The (Burning) Case for a Green New Deal*," Common Dreams, September 17, 2019, www.commondreams.org/views/2019/09/17/danger-inspiration-review-fire-burning-case-green-new-deal.

36. Wendell Berry, "Why I Am Not Going to Buy a Computer," in *What Are People For?* (San Francisco: North Point Press, 1990), 170–77.

37. George Orwell, "Rudyard Kipling" (1942), http://orwell.ru/library/reviews/kipling/english/e_rkip.

38. Lorraine Boissoneault, "Are Humans to Blame for the Disappearance of Earth's Fantastic Beasts?," *Smithsonian Magazine*, July 31, 2017, www.smithsonianmag.com/science-nature/what-happened-worlds-most-enormous-animals-180964255/#J8YG1bdpIIz0QMMm.99.

39. David R. Montgomery, *Dirt: The Erosion of Civilizations*, 2nd ed. (Berkeley: University of California Press, 2012).

40. James C. Scott, *Against the Grain: A Deep History of the Earliest States* (New Haven, CT: Yale University Press, 2017).

41. Heather Pringle, "The Ancient Roots of the 1%," *Science* 344, no. 6186 (2014): 822–25, www.science.org/lookup/doi/10.1126/science.344.6186.822; Manvir Singh, "Beyond the !Kung," *Aeon*, February 8, 2021, https://aeon.co/essays/not-all-early-human-societies-were-small-scale-egalitarian-bands.

42. Sterelny, "How Equality Slipped Away."

Chapter Two. Four Hard Questions

1. "A Brief Explanation of the Overton Window," Mackinac Center for Public Policy, www.mackinac.org/OvertonWindow.

2. Marie von Ebner-Eschenbach, *Aphorisms* (Riverside, CA: Ariadne Press, 1994), 28.

3. Ronald Bailey, *The End of Doom* (New York: St. Martin's Press, 2015), 9.

4. Paul Ehrlich, *The Population Bomb* (New York: Sierra Club/Ballantine Books, 1968), xi.

5. Julian Simon, *The Ultimate Resource 2* (Princeton, NJ: Princeton University Press, 1996), ch. 3, www.juliansimon.com/writings/Ultimate_Resource/.

6. Bruce Pengra, "One Planet, How Many People? A Review of Earth's Carrying Capacity," UNEP Global Environmental Alert Service (2012), https://na.unep.net/geas/archive/pdfs/geas_jun_12_carrying_capacity.pdf.

7. See Richard Heinberg, "Ted Nordhaus Is Wrong: We Are Exceeding Earth's Carrying Capacity," *Undark*, July 26, 2018, https://undark.org/2018/07/26/ted-nordhaus-carrying-capacity-ecology/; Ted Nordhaus, "The Earth's Carrying Capacity for Human Life Is not Fixed," *Aeon*, July 5, 2018, https://aeon.co/ideas/the-earths-carrying-capacity-for-human-life-is-not-fixed.

8. William R. Catton Jr., *Overshoot: The Ecological Basis of Revolutionary Change* (Urbana: University of Illinois Press, 1980).

9. For an introduction to the issue, see Diana Coole, "Population, Environmental Discourse, and Sustainability," in *The Oxford Handbook of Environmental Political Theory*, eds. Teena Gabrielson, Cheryl Hall, John M. Meyer, and David Schlosberg (Oxford: Oxford University Press, 2016), 274–88.

10. Sarah Kaplan, "It's Wrong to Blame 'Overpopulation' for Climate Change," *Washington Post*, May 25, 2021, www.washingtonpost.com/climate-solutions/2021/05/25/slowing-population-growth-environment/.

11. Carly Goodman, "The Shadowy Network Shaping Trump's Anti-immigration Policies," *Washington Post*, September 27, 2018, www.washingtonpost.com/outlook/2018/09/27/shadowy-network-shaping-trumps-anti-immigration-policies/; Peter Beinart, "White Nationalists Discover the Environment," *The Atlantic*, August 5, 2019, www.theatlantic.com/ideas/archive/2019/08/white-nationalists-discover-the-environment/595489/.

12. Caitlin Fendley, "Eugenics Is Trending; That's a Problem," *Washington Post*, February 17, 2020, www.washingtonpost.com/outlook/2020/02/17/eugenics-is-trending-thats-problem/.

13. Sam Knights, "The Climate Movement Must Be Ready to Challenge Rising Right-Wing Environmentalism," *Jacobin*, November 16, 2020, https://jacobinmag.com/2020/11/climate-change-right-wing-environmentalism-alt -right-eco-fascism/; Max Ajl, "Eco-Fascisms and Eco-Socialisms," Verso Blog, August 12, 2019, www.versobooks.com/blogs/4404-eco-fascisms-and-eco-socialisms.

14. Eileen Crist, "Decoupling the Global Population Problem from Immigration Issues," *Ecological Citizen* 2, no. 2 (2019): 150, www.ecologicalcitizen.net/article.php?t=decoupling-global-population-problem-immigration-issues.

15. Daphne H. Liu and Adrian E. Raftery, "How Do Education and Family Planning Accelerate Fertility Decline?," *Population & Development Review* 46, no. 3 (2020): 409–41.

16. Mara Hvistendahl, "Analysis of China's One-Child Policy Sparks Uproar," *Science*, October 18, 2017, www.sciencemag.org/news/2017/10/analysis-china-s-one-child-policy-sparks-uproar.

17. "Life Expectancy," *Scientific American*, May 13, 2002, www.scientificamerican.com/article/life-expectancy/.

18. James Ridgeway, "Meet the Real Death Panels," *Mother Jones*, July–August 2010, www.motherjones.com/politics/2010/07/health-care-rationing-death-panels/.

19. Steven Johnson, "How Humanity Gave Itself an Extra Life," *New York Times*, April 27, 2021, www.nytimes.com/2021/04/27/magazine/global-life-span

.html. Excerpted from Steven Johnson, *Extra Life: A Short History of Living Longer* (New York: Riverhead Books, 2021).

20. Damien Cave, Emma Bubola, and Choe Sang-Hun, "Long Slide Looms for World Population, with Sweeping Ramifications," *New York Times*, May 22, 2021, www.nytimes.com/2021/05/22/world/global-population-shrinking.html.

21. Robert Jensen, *The Restless and Relentless Mind of Wes Jackson: Searching for Sustainability* (Lawrence: University Press of Kansas, 2021), 61–63.

22. William E. Rees, "Ecological Economics for Humanity's Plague Phase," *Ecological Economics* 169 (2020): 8, www.sciencedirect.com/science/article/abs/pii/S0921800919310699?via%3Dihub.

23. Christopher Tucker, *A Planet of 3 Billion: Mapping Humanity's Long History of Ecological Destruction and Finding Our Way to a Resilient Future: A Global Citizen's Guide to Saving the Planet* (Alexandria, VA: Atlas Observatory Press, 2019).

24. Alexandra Minna Stern, "White Nationalists' Extreme Solution to the Coming Environmental Apocalypse," The Conversation, August 22, 2019, https://theconversation.com/white-nationalists-extreme-solution-to-the-coming-environmental-apocalypse-121532.

25. Jackson likes to say there are only two things wrong with that title: the words *Man* and *Environment*. The book was published when it was still common to use "man" to mean "humans" or "people." And "and the environment" should be replaced by "in ecosystems" or "as part of the ecosphere" to avoid reinforcing the idea that humans are somehow separate from the environment.

26. Wes Jackson, *Man and the Environment*, 3rd ed. (Dubuque, IA: W. C. Brown, 1979).

27. Paul R. Ehrlich and John P. Holdren, "Impact of Population Growth," *Science*, 171, no. 3977 (1971): 1212–17, https://science.sciencemag.org/content/171/3977/1212.

28. Donella H. Meadows, Dennis L. Meadows, Jørgen Randers, and William W. Behrens III, *The Limits to Growth: A Report for the Club of Rome's Project on the Predicament of Mankind* (New York: Signet Books, 1972).

29. Gaya Herrington, "Update to Limits to Growth: Comparing the World3 Model with Empirical Data," *Journal of Industrial Ecology* 25, no. 3 (2021): 614–26, https://onlinelibrary.wiley.com/doi/abs/10.1111/jiec.13084; Graham Turner and Cathy Alexander, "Limits to Growth Was Right. New Research Shows We're Nearing Collapse," *The Guardian*, September 1, 2014, www.theguardian.com/commentisfree/2014/sep/02/limits-to-growth-was-right-new-research-shows-were-nearing-collapse.

30. Catton, *Overshoot*.

31. Mathis Wackernagel and William Rees, *Our Ecological Footprint: Reducing Human Impact on the Earth* (Gabriola Island, BC: New Society Publishers, 1996).

32. Richard Heinberg, *Power: Limits and Prospects for Human Survival* (Gabriola Island, BC: New Society Publishers, 2021).

33. "Founded in 2003, Post Carbon Institute's mission is to lead the transition to a more resilient, equitable, and sustainable world by providing individuals and communities with the resources needed to understand and respond to the interrelated ecological, economic, energy, and equity crises of the 21st century." www.postcarbon.org/about-us/.

34. Corey J. A. Bradshaw et al., "Underestimating the Challenges of Avoiding a Ghastly Future," *Frontiers in Conservation Science* 1, no. 615419 (2021): 3, www.frontiersin.org/articles/10.3389/fcosc.2020.615419/full.

35. Wes Jackson, *Altars of Unhewn Stone: Science and the Earth* (San Francisco: North Point Press, 1987), 150; Jensen, *The Restless and Relentless Mind of Wes Jackson*, 31–33.

36. Subrena E. Smith, "Is Evolutionary Psychology Possible?," *Biological Theory* 15 (2020): 39–49, https://link.springer.com/article/10.1007/s13752-019-00336-4.

37. E. O. Wilson, *Sociobiology: The New Synthesis, Twenty-Fifth Anniversary Edition* (Cambridge, MA: Harvard University Press, 2000), 4.

38. Anthony Gottlieb, "It Ain't Necessarily So: How Much Do Evolutionary Stories Reveal about the Mind?," *New Yorker*, September 10, 2012, www.newyorker.com/magazine/2012/09/17/it-aint-necessarily-so.

39. Robin Dunbar, *The Human Story* (London: Faber and Faber, 2004).

40. Malcolm Gladwell, "The Cellular Church: How Rick Warren's Congregation Grew," *New Yorker*, September 5, 2005, www.newyorker.com/magazine/2005/09/12/the-cellular-church.

41. "How to Form an Affinity Group; The Essential Building Block of Anarchist Organization," CrimethInc. (2017), https://crimethinc.com/2017/02/06/how-to-form-an-affinity-group-the-essential-building-block-of-anarchist-organization.

42. David Stasavage, "Lessons from all Democracies," *Aeon*, March 9, 2021, https://aeon.co/essays/democracy-is-common-and-robust-historically-and-across-the-globe.

43. Alexa Clay, "Utopia Inc," *Aeon*, February 28, 2017, https://aeon.co/essays/like-start-ups-most-intentional-communities-fail-why.

44. As people do on their own, or with help from organizations such as the Institute for Local Self-Reliance. See https://ilsr.org/about-the-institute-for-local-self-reliance/.

45. The term comes from David Orr, "Technological Fundamentalism," *Conservation Biology* 8, no. 2 (June 1994): 335–37. See also Jensen, *The Restless and Relentless Mind of Wes Jackson*, 80–82.

46. Jensen, *The Restless and Relentless Mind of Wes Jackson*, 82–84.

47. Lisa Friedman, "Biden Administration Makes First Major Move to Regulate Greenhouse Gases," *New York Times*, September 23, 2021, www.nytimes.com/2021/09/23/climate/hydrofluorocarbons-hfc-climate-change.html.

48. O. B. Tsvetkov, Yu A. Laptev, A. V. Sharkov, V. V. Mitropov, and A. V. Fedorov, "Alternative Refrigerants with Low Global Warming Potential for Refrigeration and Air-conditioning Industries," *2020 IOP Conference Series: Materials Science and Engineering*, https://iopscience.iop.org/article/10.1088/1757-899X/905/1/012070.

49. Stan Cox, "100 Percent Wishful Thinking: The Green-Energy Cornucopia," Green Social Thought, September 7, 2017, www.greensocialthought.org/content/100-percent-wishful-thinking-green-energy-cornucopia.

50. Michael Shellenberger, *Apocalypse Never: Why Environmental Alarmism Hurts Us All* (New York: HarperCollins, 2020).

51. Elizabeth Kolbert, *Under a White Sky: The Nature of the Future* (New York: Crown, 2021).

52. Naomi Klein, *This Changes Everything: Capitalism vs. the Climate* (New York: Simon & Schuster, 2014), ch. 8, "Dimming the Sun: The Solution to Pollution Is . . . Pollution."

53. David Keith, "What's the Least Bad Way to Cool the Planet?," *New York Times*, October 1, 2021, www.nytimes.com/2021/10/01/opinion/climate-change-geoengineering.html.

54. Mark Bittman, *Animal, Vegetable, Junk: A History of Food, from Sustainable to Suicidal* (New York: Houghton Mifflin Harcourt, 2021), ch. 12, "The So-Called Green Revolution"; Raj Patel, "The Long Green Revolution," *Journal of Peasant Studies* 40, no. 1 (2012): 1–63, www.tandfonline.com/doi/full/10.1080/03066150.2012.719224.

55. Vaclav Smil, *Enriching the Earth: Fritz Haber, Carl Bosch, and the Transformation of World Food Production* (Cambridge, MA: MIT Press, 2000).

56. Hannah Ritchie, "How Many People Does Synthetic Fertilizer Feed?," Our World in Data, November 7, 2017, https://ourworldindata.org/how-many-people-does-synthetic-fertilizer-feed.

57. National Research Council, *Lessons Learned from the Fukushima Nuclear Accident for Improving Safety of U.S. Nuclear Plants* (Washington, DC: National Academies Press, 2014), www.ncbi.nlm.nih.gov/books/NBK253923/.

58. Charles Perrow, *Normal Accidents: Living with High Risk Technologies*, updated ed. (Princeton, NJ: Princeton University Press, 1999).

59. Stan Cox, *The Path to a Livable Future: A New Politics to Fight Climate Change, Racism, and the Next Pandemic* (San Francisco: City Lights, 2021).

60. Bill Vitek and Wes Jackson, eds., *The Virtues of Ignorance: Complexity, Sustainability, and the Limits of Knowledge* (Lexington: University Press of Kentucky, 2008); Jensen, *The Restless and Relentless Mind of Wes Jackson*, 77–80.

61. For an *extremely* different account, see John Asafu-Adjaye et al., "An Ecomodernist Manifesto," 2015, www.ecomodernism.org/.

62. Ian Tattersall, *Masters of the Planet: The Search for Our Human Origins* (New York: Palgrave Macmillan, 2012), 227.

63. Swiss Re Institute, "A Fifth of Countries Worldwide at Risk from Ecosystem Collapse as Biodiversity Declines, Reveals Pioneering Swiss Re Index" (2020), www.swissre.com/media/news-releases/nr-20200923-biodiversity-and-ecosystems-services.html.

64. Swiss Re Institute, "The Economics of Climate Change: No Action not an Option" (2021), www.swissre.com/institute/research/topics-and-risk-dialogues/climate-and-natural-catastrophe-risk/expertise-publication-economics-of-climate-change.html.

65. Deloitte Center for Financial Services, "Climate Risk: Regulators Sharpen Their Focus" (2019), 15, www2.deloitte.com/content/dam/Deloitte/us/Documents/financial-services/us-fsi-climate-risk-regulators-sharpen-their-focus.pdf.

66. Climate-Related Market Risk Subcommittee, "Managing Climate Risk in the U.S. Financial System" (U.S. Commodity Futures Trading Commission, Market Risk Advisory Committee, Washington, DC, 2020), i.

67. Daniel R. Coats, Director of National Intelligence, "Worldwide Threat Assessment of the U.S. Intelligence Community" (2019), 23, www.dni.gov/files/ODNI/documents/2019-ATA-SFR-SSCI.pdf.

Chapter Three. We Are All Apocaplyptic Now

1. As well as Jensen can recollect, his first published use of the phrase was in the article, "Rationally Speaking, We Are All Apocalyptic Now," Truthout, February 8, 2013, https://truthout.org/articles/rationally-speaking-we-are-all-apocalyptic-now/. That was followed by a lecture, "We Are All Apocalyptic Now: Moral Responsibilities in Crisis Times," delivered at the First Unitarian Universalist Church Public Affairs Forum, Austin, TX, February 24, 2013, www.youtube.com/watch?v=aqgigYmr86Y. Jensen self-published a short book, or long pamphlet, later that

year: *We Are All Apocalyptic Now: On the Responsibilities of Teaching, Preaching, Reporting, Writing, and Speaking Out* (n.p.: CreateSpace, 2013). We draw on that work for this chapter.

2. Quoted in Susan George, *Whose Crisis, Whose Future: Towards a Greener, Fairer, Richer World* (Cambridge: Polity, 2010), 169–70.

3. "The Economy: We Are All Keynesians Now," *Time*, December 31, 1965, http://content.time.com/time/magazine/article/0,9171,842353,00.html.

4. James Gleick, *Genius: The Life and Science of Richard Feynman* (New York: Pantheon, 1992). www.around.com/genius.html.

5. Walter Brueggemann, *The Prophetic Imagination*, 2nd ed. (Minneapolis, MN: Fortress Press, 2001), 41.

6. Ibid., 40.

7. Abraham J. Heschel, *The Prophets* (New York: HarperCollins, 2001), 19; emphasis added.

8. Walter Brueggemann, *The Practice of Prophetic Imagination* (Minneapolis, MN: Fortress Press, 2012), 69.

9. "The Earth will be completely laid waste and totally plundered. The LORD has spoken this word. The Earth dries up and withers, the world languishes and withers, the heavens languish with the Earth" (Isaiah 24:3–4 NIV).

10. James Baldwin, "As Much Truth as One Can Bear," in *The Cross of Redemption: Uncollected Writings*, ed. Randall Kenan (New York: Pantheon, 2010), 34, 29.

11. Shoba Sreenivasan and Linda E. Weinberger, "Fear Appeals: An Approach Used to Change Our Attitudes and Behaviors, *Psychology Today*, September 18, 2018, www.psychologytoday.com/us/blog/emotional-nourishment/201809/fear-appeals.

12. Barbara Ehrenreich, *Bright-sided: How Positive Thinking Is Undermining America* (New York: Picador, 2010).

13. Future Earth, "Risks Perceptions Report 2020: First Edition," https://futureearth.org/wp-content/uploads/2020/02/RPR_2020_Report.pdf.

14. Thomas Wiedmann, Manfred Lenzen, Lorenz T. Keyßer, and Julia K. Steinberger, "Scientists' Warning on Affluence," *Nature Communications* 11, no. 3107 (2020), www.nature.com/articles/s41467-020-16941-y.

15. Pablo Servigne and Raphaël Stevens, *How Everything Can Collapse: A Manual for Our Times*, trans. Andrew Brown (Cambridge: Polity, 2020).

16. See the work of scholars associated with the "Global Systemic Risk" Research Community, Princeton Institute for International and Regional Studies, Princeton University. https://risk.princeton.edu/.

17. Pablo Servigne, Raphaël Stevens, and Gauthier Chapelle, *Another End of the World Is Possible*, trans. Geoffrey Samuel (Cambridge: Polity, 2021).

18. Laura Spinney, "'Humans Weren't Always Here. We Could Disappear': Meet the Collapsologists," *The Guardian*, October 11, 2020, www.theguardian.com/world/2020/oct/11/humans-werent-always-here-we-could-disappear-meet-the-collapsologists.

19. Ben Ehrenreich, "How Do You Know When Society Is About to Fall Apart?," *New York Times*, November 4, 2020, www.nytimes.com/2020/11/04/magazine/societal-collapse.html.

20. Jem Bendell, "Deep Adaptation: A Map for Navigating the Climate Tragedy" (2019), https://jembendell.com/2019/05/15/deep-adaptation-versions/.

21. Jonah Engel Bromwich, "The Darkest Timeline," *New York Times*, December 26, 2020, www.nytimes.com/2020/12/26/style/climate-change-deep-adaptation.html.

22. Jared Diamond, *Collapse: How Societies Choose to Fail or Succeed* (New York: Viking Press, 2005), 3.

23. Joseph A. Tainter, "Collapse, Sustainability, and the Environment: How Authors Choose to Fail or Succeed," *Reviews in Anthropology* 37 (2008): 342–71, www.tandfonline.com/doi/abs/10.1080/00938150802398677.

24. Joseph A. Tainter, *The Collapse of Complex Societies* (Cambridge: Cambridge University Press, 1988).

25. Ibid., 4.

26. Peter Turchin, "America in November 2020: A Structural-Demographic View from Alpha Centauri," PeterTurchin.com, November 1, 2020, http://peterturchin.com/cliodynamica/america-in-november-2020-a-structural-demographic-view-from-alpha-centauri/.

27. For details, see Miguel A. Centeno, Manish Nag, Thayer S. Patterson, Andrew Shaver, and A. Jason Windawi, "The Emergence of Global Systemic Risk," *Annual Review of Sociology* 41 (2015): 65–85, www.annualreviews.org/doi/10.1146/annurev-soc-073014-112317.

28. Raj Patel, "Agroecology Is the Solution to World Hunger," *Scientific American*, September 22, 2021, www.scientificamerican.com/article/agroecology-is-the-solution-to-world-hunger/.

29. Oliver Balch, "The Curse of 'White Oil': Electric Vehicles' Dirty Secret," *The Guardian*, December 8, 2020, www.theguardian.com/news/2020/dec/08/the-curse-of-white-oil-electric-vehicles-dirty-secret-lithium; Patrick Moriarty and Stephen Jia Wang, "Can Electric Vehicles Deliver Energy and Carbon Reductions?," *Energy Procedia* 105 (May 2017): 2983–88, www.sciencedirect.com/science/article/pii/S1876610217307762; Troy R. Hawkins, Bhawna Singh, Guillaume

Majeau-Bettez, and Anders Hammer Strømman, "Comparative Environmental Life Cycle Assessment of Conventional and Electric Vehicles," *Journal of Industrial Ecology* 17, no. 1 (February 2013): 53–64, https://onlinelibrary.wiley.com/doi/10.1111/j.1530-9290.2012.00532.x.

30. Lao Tzu, *Tao Te Ching*, ch. 80. Accessed at https://taoism.net/tao-te-ching-online-translation/.

31. Mark O'Connell, *Notes from an Apocalypse: A Personal Journey to the End of the World and Back* (New York: Doubleday, 2020).

32. See the work of the Transition Network, whose motto is, "A movement of communities coming together to reimagine and rebuild our world." https://transitionnetwork.org/.

33. Wes Jackson and Robert Jensen, "Let's Get 'Creaturely': A New Worldview Can Help Us Face Ecological Crises," in *The Perennial Turn: Contemporary Essays from the Field*, ed. Bill Vitek (Middlebury, VT: New Perennials Publishing, 2020), 130–37.

34. John Gorka, "Old Future," from the CD *Old Futures Gone* (Red House Records, 2003).

Chapter Four. A Saving Remnant

1. "The Käte Hamburger Centre for Apocalyptic and Post-Apocalyptic Studies (CAPAS) at Heidelberg University focuses on the effects of catastrophes and end-time scenarios on societies, individuals and environments." www.capas.uni-heidelberg.de/index.en.html.

2. Lester V. Meyer, "Remnant," in *Anchor Bible Dictionary*, vol. 5, ed. David Noel Freedman (New York: Doubleday, 1992), 669.

3. Lawrence O. Richards, *New International Encyclopedia of Bible Words* (Grand Rapids, MI: Zondervan, 1999), 521.

4. Cedric B. Cowing, *The Saving Remnant: Religion and the Settling of New England* (Urbana: University of Illinois Press, 1995), 1.

5. Herbert Agar, *The Saving Remnant: An Account of Jewish Survival* (New York: Viking, 1960), 6.

6. W. E. B. Du Bois, "The Talented Tenth," in *The Negro Problem: A Series of Articles by Representative American Negroes of Today* (New York: James Pott and Co., 1903), 43; available at https://archive.org/details/negroproblemseri00washrich/mode/2up.

7. Ibid., 44.

8. Martin Duberman, *A Saving Remnant: The Radical Lives of Barbara Deming and David McReynolds* (New York: New Press, 2011), xi.

9. Billy Ireland Cartoon Library and Museum, The Ohio State University, https://library.osu.edu/site/40stories/2020/01/05/we-have-met-the-enemy/.

10. Marshall Sahlins, *Stone Age Economics*, 2nd ed. (New York: Routledge, 2004), 1–2.

11. Wallace Stegner, *Beyond the Hundredth Meridian: John Wesley Powell and the Second Opening of the West* (New York: Houghton Mifflin, 1953), 256.

12. Cary Telander Fortin and Kyle Louise Quilici, *New Minimalism: Decluttering and Design for Sustainable, Intentional Living* (Seattle, WA: Sasquatch Books, 2018).

13. The popularity of the 2019 Netflix series *Tidying Up with Marie Kondo* suggests that many people find such advice helpful. https://learn.konmari.com/.

14. Marie Kondo, *The Life-Changing Magic of Tidying Up: The Japanese Art of Decluttering and Organizing* (Emeryville, CA: Ten Speed Press, 2014).

15. Warren Johnson, *Muddling toward Frugality: A New Social Logic for a Sustainable World*, rev. ed. (Norwalk, CT: Easton Studio Press, 2010).

16. Duane Elgin, *Voluntary Simplicity: Toward a Way of Life That Is Outwardly Simple, Inwardly Rich*, 2nd rev. ed. (New York: HarperCollins, 2010).

17. Jia Tolentino, "The Pitfalls and the Potential of the New Minimalism," *New Yorker*, February 3, 2020, 73, www.newyorker.com/magazine/2020/02/03/the-pitfalls-and-the-potential-of-the-new-minimalism.

18. Guy Champniss, Hugh N. Wilson, and Emma K. Macdonald, "Why Your Customers' Social Identities Matter," *Harvard Business Review* (January–February 2015), https://hbr.org/2015/01/why-your-customers-social-identities-matter.

19. Justin Lewis, *Beyond Consumer Capitalism: Media and the Limits to Imagination* (Cambridge: Polity, 2013).

20. Julian Brave Noisecat, "Apocalypse Then and Now," *Columbia Journalism Review* (Winter 2020), www.cjr.org/special_report/apocalypse-then-and-now.php.

21. Judy Dow, "Going through the Narrows," *Potash Hill* (Spring 2019): 7.

22. Andrew Fletcher, "An Account of a Conversation concerning a Right Regulation of Governments for the Common Good of Mankind," in *The Political Works of Andrew Fletcher* (London: James Bettenham Press, 1737), 372.

23. Jensen would offer as an example his wife Eliza Gilkyson's CD *Beautiful World* (Red House Records, 2008).

24. Judy Hart Angelo and Gary Portnoy, "Where Everybody Knows Your Name: Theme from *Cheers*" (1982), Vocal Popular Sheet Music Collection, Score 5142, https://digitalcommons.library.umaine.edu/mmb-vp-copyright/5142.

25. Derrick Jensen, Lierre Keith, and Max Wilbert, *Bright Green Lies: How the Environmental Movement Lost Its Way and What We Can Do about It* (Rhinebeck, NY: Monkfish Book Publishing, 2021).

26. David Graeber and David Wengrow, *The Dawn of Everything: A New History of Humanity* (New York: Farrar, Straus and Giroux, 2021). This book does an excellent job of highlighting the varied ways that societies throughout history have fostered human freedom. But the authors fail to acknowledge the roles of geography and biology in shaping societies and individuals, and the book's challenge to the centrality of agriculture in changing the course of human history falls flat.

27. Robert Jensen, *The Restless and Relentless Mind of Wes Jackson: Searching for Sustainability* (Lawrence: University Press of Kansas, 2021), 9–10.

Chapter Five. Ecospheric Grace

1. Robert Jensen, *All My Bones Shake: Seeking a Progressive Path to the Prophetic Voice* (Berkeley, CA: Soft Skull, 2009).

2. The Epistle to the Ephesians is attributed to Paul, but most scholars consider it one of the "deutero-Pauline epistles" that likely was written by his followers after his death.

3. We're using the male pronoun here, following the conventional usage, but we think people should ponder why that patriarchal tradition is so easily embraced by so many. See Robert Jensen, "Does God Have A Gender?," *MS Magazine*, April 23, 2013, https://msmagazine.com/2013/04/23/does-god-have-a-gender/.

4. Wes Jackson, "The Failure of Stewardship," in *New Roots for Agriculture* (Lincoln: University of Nebraska Press, 1980), 11–13.

5. Stephen Jay Gould, "Planet of the Bacteria," *Washington Post*, November 13, 1996, www.washingtonpost.com/archive/1996/11/13/planet-of-the-bacteria/6fb60f1d-e6fe-471e-8a0f-4cfa9373772c/. Adapted from Gould's *Full House* (New York: Harmony Books, 1996).

6. Sigmund Freud, *The Future of an Illusion* (New York: Anchor Books, 1964), 27.

Conclusions

1. Gerda Lerner, *The Creation of Patriarchy* (New York: Oxford University Press, 1986).

2. The Who, "Won't Get Fooled Again," from the album *Who's Next* (Track Records, 1971).

3. Martin Luther King Jr., "Where Do We Go from Here?," Annual Report to the Southern Christian Leadership Conference (1967), www.stanford.edu/group/King/publications/speeches/Where_do_we_go_from_here.html.

4. Robert Jensen, "Power Basics: Political but More than Politics," in *Arguing for Our Lives: A User's Guide to Constructive Dialogue* (San Francisco: City Lights, 2013), 29–43.

5. Ian Tattersall, *Masters of the Planet: The Search for Our Human Origins* (New York: Palgrave Macmillan, 2012), xiv.

6. Melanie Challenger, *How to Be Animal: A New History of What It Means to Be Human* (New York: Penguin Books, 2021), 1.

7. There is a school of thought in academic philosophy called "animalism" that tracks with some of our understanding. But like so much of academic philosophy, we're not sure we understand it all. See Stephan Blatti, "Animalism," in *The Stanford Encyclopedia of Philosophy*, ed. Edward N. Zalta (Fall 2020 Edition), https://plato.stanford.edu/archives/fall2020/entries/animalism/.

8. David Hume, *A Treatise of Human Nature*, bk. 2, pt. 3, sec. 3, "Of the Influencing Motives of the Will" (1739), https://davidhume.org/texts/t/2/3/3.

9. *Abe Osheroff: One Foot in the Grave, the Other Still Dancing*, dir. Nadeem Uddin (Northampton, MA: Media Education Foundation, 2009).

10. Marie von Ebner-Eschenbach, *Aphorisms* (Riverside, CA: Ariadne Press, 1994), 28.

11. Wes Jackson, *Hogs Are Up: Stories of the Land, with Digressions* (Lawrence: University Press of Kansas, 2021), 157–64.

12. Robert Jensen, *Plain Radical: Living, Loving, and Learning to Leave the Planet Gracefully* (Berkeley, CA: Counterpoint/Soft Skull, 2015).

13. Robert Jensen, *The End of Patriarchy: Radical Feminism for Men* (North Melbourne, Australia: Spinifex Press, 2017).

14. Louis Dumont used the term *Homo hierarchicus* in his 1966 book, *Homo Hierarchicus: The Caste System and Its Implications*, trans. Mark Sainsbury, Louis Dumont, and Basia Gulati (Chicago: University of Chicago Press, 1980).

15. William R. Catton Jr. coined this term in *Overshoot: The Ecological Basis of Revolutionary Change* (Urbana: University of Illinois Press, 1980).

16. Yves Gingras offered that term in *Éloge de l'homo techno-logicus*, roughly translated as *In Praise of Homo Techno-Logicus* (Saint-Laurent, QC: Fides, 2005).

17. Joanna Macy and Chris Johnstone, *Active Hope: How to Face the Mess We're in without Going Crazy* (Novato, CA: New World Library, 2012). Hannah

Arendt argued that hope too often leads to passivity; see Arendt's *The Human Condition*, 2nd ed. (Chicago: University of Chicago Press, 1998).

18. See also David Korten, *The Great Turning: From Empire to Earth Community* (San Francisco: Berrett-Koehler, 2006); Craig Schindler and Gary Lapid, *The Great Turning: Personal Peace, Global Victory* (Rochester, VT: Bear & Co., 1989).

19. Jane Goodall and Douglas Abrams, with Gail Hudson, *The Book of Hope: A Survival Guide for Trying Times* (New York: Celadon Books/Macmillan, 2021), 8.

20. Octavia E. Butler, *Parable of the Sower* (New York: Four Walls, Eight Windows, 1993) and *Parable of the Talents* (New York: Seven Stories Press, 1998). Butler planned a final book in the trilogy, tentatively titled *Parable of the Trickster*, which she did not finish before her death in 2006.

21. Robert Jensen, *Getting Off: Pornography and the End of Masculinity* (Boston: South End Press, 2007). (Out of print, available at https://robertwjensen.org/books/getting-off/.)

22. Robert Jensen, *Citizens of the Empire: The Struggle to Claim Our Humanity* (San Francisco: City Lights, 2004).

23. Wendell Berry, "Sabbaths 2007, VI," in *Leavings: Poems* (Berkeley, CA: Counterpoint, 2010), 91–93.

24. Fyodor Dostoyevsky, *The Brothers Karamazov*, trans. Constance Garnett (New York: Barnes and Noble Books, 2004), 61; emphasis added.

INDEX

WES JACKSON is cofounder and president emeritus of The Land Institute in Salina, Kansas. A 1992 MacArthur Fellow, he is the author and co-author of numerous books, including *Hogs Are Up: Stories of the Land, with Digressions and New Roots for Agriculture*.

ROBERT JENSEN is professor emeritus in the School of Journalism at the University of Texas at Austin. He is the author of many books including *The Restless and Relentless Mind of Wes Jackson: Searching for Sustainability* and *Plain Radical: Living, Loving, and Learning to Leave the Planet Gracefully*.

CPSIA information can be obtained
at www.ICGtesting.com
Printed in the USA
LVHW081344200722
723943LV00009B/400